T0140658

PUBLIKATIONEN
DER BAYERISCHEN AMERIKA-AKADEMIE
Band 12

PUBLICATIONS
OF THE BAVARIAN AMERICAN ACADEMY
Volume 12

SERIES EDITOR
Board of Directors of the Bavarian American Academy

NOTE ON THE EDITORS

Christof Mauch is Professor of American History and Transatlantic Relations, Director of the Rachel Carson Center for Enviromnment and Society at LMU Munich and Member of the Bavarian American Academy.

Sylvia Mayer is Professor of American Studies and Anglophone Literatures and Cultures at the University of Bayreuth and Member of the Bavarian American Academy.

American Environments:

Climate–Cultures–Catastrophe

Edited by
CHRISTOF MAUCH
SYLVIA MAYER

Universitätsverlag
WINTER
Heidelberg

Bibliografische Information der Deutschen Nationalbibliothek

Die Deutsche Nationalbibliothek verzeichnet diese Publikation
in der Deutschen Nationalbibliografie;
detaillierte bibliografische Daten sind im Internet
über *http://dnb.d-nb.de* abrufbar.

Rachel
Carson
Center

This publication was generously supported
by the Rachel Carson Center for Environment and Society
at LMU Munich.

COVER ILLUSTRATION

California Highway 139 © jcookfisher
http://www.flickr.com/photos/jcookfisher/5671298974/

ISBN 978-3-8253-6005-4

© 2012 Universitätsverlag Winter GmbH Heidelberg
Imprimé en Allemagne · Printed in Germany
Gesamtherstellung: Memminger MedienCentrum, 87700 Memmingen

Gedruckt auf umweltfreundlichem, chlorfrei gebleichtem
und alterungsbeständigem Papier

Den Verlag erreichen Sie im Internet unter:
www.winter-verlag.de

Preface

As with all previous volumes of the Bavarian American Academy (BAA) monograph series the following collection of essays, titled "American Environments: Climate—Cultures—Catastrophe," would not have materialized without the generous support of several institutions and the personal commitment of those who have been involved either as contributors, editors, or proofreaders. We want to thank, above all, the Rachel Carson Center for Environment and Society at LMU Munich and its director, Prof. Christof Mauch, for co-organizing and also co-sponsoring the BAA's 2010 annual conference "Green Cultures—Environmental Knowledge, Climate, and Catastrophe," from which the essays assembled here are taken. Special thanks also go to Prof. Sylvia Mayer, Chair of American Studies at the University of Bayreuth and, with Christof Mauch, one of the editors of the volume. Dr. Meike Zwingenberger, the BAA's executive director, and Jasmin Falk, M.A., our office manager, have been the driving force behind both the physical conference at the Amerika Haus in Munich and the ensuing publication of its proceedings. Jasmin Falk, in particular, has provided meticulous editorial and technical support.

Munich, January 2012

Klaus Benesch
Director, Bavarian American Academy

Table of Contents

Introduction

Christof Mauch and Sylvia Mayer

In the natural sciences, the prominent position afforded to environmental issues is long established and unquestioned. Anybody who wants to explain the mechanisms of an earthquake or phenomena such as climate change or desertification needs scientific evidence and analysis. But once we try to understand the meaning of environmental change, the humanities come into play: because all environmental challenges originate in human interaction with nonhuman nature they must be located within larger historical, social, and cultural contexts. For about three decades now, the emerging field of the environmental humanities has been addressing the current global environmental crisis from the perspectives provided by different disciplines and in ways that significantly complement scientific efforts. The environmental humanities help us understand the connection between cultural values, individual experience, and human decision on the one side, and environmental change on the other. If we want to understand the complex relationship between nature and culture, the double-edgedness of human intervention in the natural environment—the destruction and conservation, the pitfalls and successes—it does not suffice to rely on scientific knowledge. Visions and emotions, for instance, but also historical narratives and fictional texts—the way we tell stories about nature—may be regarded as performing a specific role in the process of environmental knowledge production. Their imaginative range provides 'cultural spaces' in which environmentally relevant ideas, values, even whole epistemologies, are explored by connecting the personal with the large-scale perspective, by linking individual lives to global issues, and by combining intellectual, emotional, and sensual experiences.

Environmental humanities as a field of thought developed in the United States before it gained a foothold elsewhere. There may be many reasons for this. One of them, we contend, is the fact that "nature"—in particular "wild" nature, or "wilderness," seemingly untouched by human activities—has played a central role in US cultural history. America was only sparsely settled until the late nineteenth century. "In the beginning all the world was America," wrote John Locke in his *Second Treatise on Government* in 1690. To him, America served as metaphor for the state of nature, for a distant and isolated continent. Ever since the beginning of European colonization, the economic transformation of the North American continent has been accompanied and in part driven by the development of influential concepts of the natural environment. US history and political culture, especially in the late eighteenth and in the nineteenth centuries, are linked to the notion of the United States as "nature's nation" (Perry Miller)—a nation whose "character" is formed by human interaction with the natural environment. At the turn of the nineteenth to the twentieth century, moreover, the American conservation and preservation movements pioneered environmentalist efforts that were based on an awareness of the fragility of specific landscapes and ecosystems and that led to the creation of the first national parks worldwide. Intensive study of the environment, both in the sciences and in the environmental humanities, has its origins in a long and influential cultural engagement with a diversity of US American landscapes.

This collection presents essays from a variety of disciplines: environmental history, cultural geography, political science, media studies, and cultural studies. They all draw on the methods and skills of the humanities and reveal aspects about the environment that have been largely ignored by natural scientists. All the essays have their roots in a conference that the Bavarian American Academy organized together with the Rachel Carson Center for Environment and Society in Munich in the summer of 2010. The conference discussed environmental knowledge production in the United States by focusing on the impact of natural catastrophes and on public debates on climate change, environmental risk perception, and environmental threats.

Putting risk and disaster center stage seemed like a good idea because of the frequency and destructive power of natural disasters in the United States. Indeed, the damage to the economy through natural catastrophes—tornadoes and earthquakes, droughts, wildfires, and hurricanes—is higher in the United States than in any other country, with the possible exception of Japan following the recent tsunami. Also, natural disasters have played a major role in the popular imagination of Americans. One of the most successful American children's books, Lyman Frank Baum's *The Wizard of Oz*, in which a girl is swept away by a tornado, is a case in point. American bookstores are filled with titles such as *Nature on the Rampage, Killer Flood, Devil Winds, Tidal Wave: No Escape, Dante's Peak, Aftershock: Earthquake in New York*, and *Storm of the Century*. Hollywood provides a steady stream of disaster movies featuring volcanoes, earthquakes, tornadoes, and even menacing asteroids. *Twister*, a 1996 movie about a group of tornado chasers, made close to five hundred million dollars at the box office. And in 2004, *The Day After Tomorrow*, a rollercoaster drama about a superstorm that devastates New York City at the start of a new Ice Age, was among the top-grossing movies worldwide. Hollywood's portrayals of the interaction between humans and nature generally have little to do with scientific reality. But the fantasies and fears that are distilled in these products of imagination play a powerful role in the general perception of the natural environment and in the production of knowledge. Besides fiction, the flooding of New Orleans in 2005, Al Gore's documentary *An Inconvenient Truth* in 2006, and the Deepwater Horizon oil spill in 2010 have aroused collective interest in catastrophes in the United States and unprecedented media coverage around the world.

This collection looks at perceptions of the natural environment, environmental change, and environmental policy in the United States. It opens with two essays that form a section titled "**Climate in America—Past and Current Perspectives.**" The essays approach the issue of climate and climate change from two different viewpoints and disciplines; one is written by an environmental historian, the other by a political scientist. In "Manifest Destiny and Manifest Disaster: Climate Perceptions and Realities in United States Territorial Expansion," Lawrence Culver addresses the historical dimension of American debates on climate and climate change by drawing attention to the transformation of the "Great American Desert" in the nineteenth and early twentieth centuries. Culver's essay shows that climate, and the issue of climate change, one of the most hotly debated issues in current national and international politics, have been sites of debate in the United States for centuries. Focusing on the region of the Great Plains—which in the eighteenth century was still known as the "Great American Desert"—he traces how both the actual transformation of the vast region and its conceptual mapping in the nineteenth and early twentieth century have relied on conceptualizations that reveal political, economic, religious, cultural, and, last but not least, scientific positions. Andreas Falke's essay poses the question "Why is the United States a Laggard in Climate Change Policy?" and focuses on the current political debate. He investigates the reasons why the US, even under the current Obama administration, which announced a much stronger commitment to the environment, must still be perceived as a laggard when it comes to climate change legislation. Starting out with a brief survey of US climate change policies since the late 1980s, Falke follows their development up to the post-2010 Congressional Election. He identifies and investigates four dimensions that account for the laggard status: public opinion, the nature of the US political process and political system, the impact of industry, and the implications of a strongly United Nations-influenced diplomacy.

The next section, "**Cultures of Ecology—Cultures of Risk**," brings together essays from the disciplines of geography, environmental history, and cultural studies. Geographer Heike Egner, in "Risk, Space, and Natural Disasters: On the Role of Nature and Space in Risk Research," starts out with theoretical remarks on contemporary risk and observation theory, on the concrete and conceptual mapping of risks via spatial indexing, and on nature as a social category. She then proceeds with two case studies: the first one focuses on a coastal community, Camp Ellis, in Saco, Maine, and highlights the problematic role of spatial indexing; the second one looks at the employment of concepts of nature in different narrations of climate change. In "Buffalo Commons: The Past, Present, and Future of an Idea," environmental historian Andrew C. Isenberg addresses the complex issue of grand-scale environmental restoration projects. Taking his cue from the idea, reinvigorated in the 1980s, of re-introducing buffalo onto the North American Great Plains, he recapitulates the history of the vast grasslands area, identifying as he does so the risks and opportunities involved in conservation efforts. It becomes obvious that the challenge conservationists face in the attempt to conceptualize any sustainable ecosystem is the "twin dynamism of history and ecology." The participation of popular culture in environmental risk discourse is targeted by cultural critic Alexa Weik von Mossner in "Facing *The Day After Tomorrow*: Filmed Disaster, Emotional Engagement, and Climate Risk Perception." Drawing on the results of recent cognitive film studies, she shows how aesthetic strategies such as melodrama and the representation of nature as spectacle engage the viewers' intellectual as well as emotional capacities in the process of transporting information about the phenomenon of (abrupt) climate change.

The last four essays focus on the issue of environmental catastrophe. Environmental historian Sherry Johnson and geographer Gordon Winder open the section **"Catastrophe—Natural Disaster and the Media"** with essays that investigate the historical significance of major earthquakes. In "The Cuban Earthquake of 1880: A Case Study from the Past with Frightening Implications for the Future," Johnson explores the historical contexts of a "forgotten" catastrophe, the Vuelta Abajo earthquake of 1880. She thereby draws attention to the significance of social memory and historiography for current disaster studies—not least when it comes to preventing future catastrophes. In "The *Los Angeles Times* Reports Japanese Earthquakes, 1923-1995," Winder shifts attention to the political and economic contexts that have an impact on disaster coverage in the media. He analyzes reports of the *Los Angeles Times* on nine earthquakes in Japan that occurred in the years 1923 to 1995 and points out both continuities and differences of representation that reflect the changing roles of the United States and Japan in the international political and economic systems. In the next essay, "Forgetting the Unforgettable: Losing Resilience in New Orleans," geographer Craig E. Colten also targets the nexus between social memory and disaster prevention. He compares how the city of New Orleans recovered from two major hurricanes, Hurricane Betsy in 1965 and Hurricane Katrina in 2005, by examining developments in land use regulations, residential architecture, and evacuation planning. Finally, taking her cue from another recent catastrophe, the Deepwater Horizon oil spill of 2010, cultural critic Stacy Alaimo, addresses the question of scientific, journalistic, and artistic representation of the oceans. In "Dispersing Disaster: The Deepwater Horizon, Ocean Conservation, and the Immateriality of Aliens," Alaimo links the striking lack of detailed knowledge about the vast ecosystem of the ocean to a lack of insight into the significance and agency of material forces and to the representation of the oceans in popular culture as alien, unreal, even immaterial spaces. Such lack of knowledge, she argues, ultimately explains the lack of a sense of human environmental accountability.

The editors would like to thank everybody who has been involved in this project. First and foremost, the directors of the Bavarian American Academy, Klaus Benesch and Meike Zwingenberger, who invited us to organize their annual conference 2010 and who suggested that we publish a collection of essays in their se-ries with Universitätsverlag Winter Heidelberg. We would also like to thank the contributors for reworking their conference papers, for sending us illustrations, and for their patience. Our very special thanks go to Katie Ritson at the Rachel Carson Center for Environment and Society at LMU Munich who has done a great job in editing, copyediting, and polishing the final versions of the essays, to Jasmin Falk of the Bavarian American Academy who has helped with the formatting of this volume and to Matthias Klestil of the University of Bayreuth and Susanne Mader for their careful proofreading. Both institutions, the Bavarian American Academy and the Rachel Carson Center have worked together very well over the last two years and we are pleased that our cooperation and our common academic interests have resulted in this collection of essays.

Manifest Destiny and Manifest Disaster: Climate Perceptions and Realities in United States Territorial Expansion

Lawrence Culver

The contentious politics of the United States in the early twenty-first century have produced debates over taxes and budgets, domestic and foreign policy, health care, and a host of other issues. One especially divisive debate concerned climate: Was the climate changing, were humans to blame, and if so, what should be done? The global scientific community's consensus that climate change was indeed underway, and was primarily due to human actions, has only strengthened since the 1990s. The political class of the United States, however, displays no such unanimity. Some claim that climate change may indeed be underway, but that proof of anthropogenic causality is unclear. Others go still further, asserting that climate change is a fiction, a hoax perpetrated by media and scientific elites. US and international oil companies have also funded research with the explicit purpose of delegitimizing climate change, or at least raising questions about it. In 2009 and 2010 alone, coal and oil companies spent an astounding half a billion dollars on political campaigns in the United States (Brune). Environmentalists see this anti-climate change stance as a product of longstanding suspicions in the US towards science, intellectuals, or "elites" of any kind. What they—and their opponents—do *not* see is that this climate debate is nothing new. Americans—and inhabitants of many other parts of the world—have in fact been discussing, debating, and arguing about climate for centuries. Their perceptions of environmental knowledge, their lack of environmental knowledge, or their wishful thinking about climate sometimes led to catastrophe. Those debates of the past, no less than those of the present, were shaped by economics and politics, by regional perspectives and national interests.

In places of European settlement and exploration, Europeans and European creoles encountered climates far different from those they had known. Most disconcerting were arid landscapes, or grasslands where trees were rare. In Australia, on the steppes of Asiatic Russia, and in the Great Plains, Pampas, and deserts of the Americas, this led to speculation and contentious debates over development, settlement, agriculture, and urbanism. Indeed, even in Europe, where landscapes and climates were long-inhabited and presumably familiar, there were debates as well, particularly over underutilized or seemingly agriculturally unproductive lands—such as the marshes and heaths of England, or wetlands and areas of sandy soil in Prussia.

By examining this climatic confusion and contestation—how individuals and nations perceived climate, and acted on those perceptions—we can uncover a far older and more complex history of climate as a cultural, political, and scientific issue, one that has played a key role in eras long preceding our own, and this history can inform our current debates concerning climate and climate policy. This essay explores one particular moment and locale in this much larger history: the nineteenth-century United States and its ideology of expansion, "Manifest Destiny;" its encounter with a roadblock to that expansion, the arid West of North America, particularly a large region that came to be called the "Great American Desert;" and a contentious debate over the development of this region that hinged on climate, and the human capacity to overcome climatic adversity through technological means, or, perhaps, even change the climate itself.

European settlers in North America, and many other places, drew upon a long experience of farming, on the agricultural folk knowledge of their cultures, on old—and sometimes ancient—religious and philosophical ideas about nature, and on the newly emergent natural sciences (Glacken). Folk belief may seem dubious by the standards of the early twenty-first century, but it had been earned by long and hard generations of agricultural work, of learning nature through labor. When explorers or settlers tried to discern climate, they were first and foremost reading landscapes. In an era before precise measurements of rainfall, humidity, or temperature, and long before regular weather and temperature records were kept, a new landscape held many clues to a region's climate. Were trees abundant, or good pasturage? Did streams and rivers seem constant? What kind of flora and fauna occupied the landscape? Did it remind them of productive agricultural landscapes they had seen or heard of in Europe? Did the climate seem healthful, and free of disease? Such questions were logical and valid. Settlers lacked modern climate science, and—no less importantly—long-term experience with these landscapes. In their native lands in Europe, they had possessed long histories of heat and cold, of floods, droughts, and blizzards, of average times to plant or harvest, of times of feast or famine. Native peoples, of course, knew all these as well, though European settlers often refused to ask or listen. This would repeatedly blind them to environmental hazards, not only in the colonial era, but also in the much more recent past (Orsi).

American Manifest Destiny Confronts the Great American Desert

One key moment in the history of perceptions of climate was the Euro-American encounter in North America with the Great Plains and the arid, high-elevation terrain of the Great Basin between the Rocky Mountains and the Sierra Nevada

Mountains. This encounter led to a cartographic creation that once appeared prominently on maps of the continent, but has now completely disappeared: the Great American Desert. Confronted with the treeless expanse that began west of the Mississippi and Missouri rivers, and stretched across ever-higher terrain all the way to the easternmost ranges of the Rocky Mountains, with deserts continuing westward beyond the mountains, white Americans thought that they had found a barrier and roadblock to their continental ambitions. Indeed, even humid areas with abundant rainfall but limited forests were initially viewed as agriculturally useless. While settlers soon discovered that the eastern prairies—what would later become states like Illinois and Iowa—were in fact productive farmland, the high plains further west seemed more daunting.

The environmental facts—a treeless landscape—were seen through political and economic prisms. Historians and average Americans alike look back on the westward expansion of the US across the continent in the nineteenth century, and see relentless purpose, a historical "destiny" made manifest. The burgeoning population and economy of the United States in this era, and the broad popularity of the ideology of Manifest Destiny—a belief in the spread of the US political and economic system, of the English language and Protestant Christianity—might in retrospect make it seem inevitable. The adherents of Manifest Destiny blithely assumed that everyone else in North America—namely, Native American Indians and Mexicans—would obligingly retreat, slowly fade away, or—in the darkest genocidal impulses of this mindset—be exterminated. Historians have ably unpacked this ideology, amply illuminating all its problematic racial, religious, and cultural aspects (Horsman). Yet even many historians still treat it and the expansion of the US as inevitable. Moreover, they ignore the ecological assumptions, which are as implicit in Manifest Destiny as its explicit racism and religious chauvinism. The entire continent itself—its climates, its soils and topography, its flora and fauna—was intended to comply with Anglo-American designs. Anything that suggested otherwise—such as the existence of the Great American Desert—was an affront to be overcome or erased.

Across the nineteenth century, some would attempt just that, asserting that the Great American Desert was no more than a passing mirage. Either that supposed aridity was merely an illusion, or technology and irrigation would defeat it, or, even more radically, humans could change an arid climate to a wet one through their own actions. Just as oil and coal industry capitalists and environmentalists and wind and solar energy developers find themselves on opposite sides of a climate divide in the early twenty-first century, political and economic interests produced a climate divide in antebellum America. In this case, the issue was in the volatile subject of slavery and its potential expansion. Many Northerners saw the Great American Desert, and the vast tracts of mountains and wilderness that lay beyond it, as ample evidence that slavery would not expand, that the nation had reached its natural limits, and that what lay beyond would best be left to Indians and wild animals.

No less a national figure than Daniel Webster, New Englander by birth, senator from Massachusetts, and secretary of state under three presidents, railed against the prospect of a federal mail route connecting the Pacific coast to the United States, let alone future annexation:

> What do we want with this vast, worthless area? This region of savages and wild beasts, of deserts, of shifting sands and whirlwinds of dust, of cactus and prairie dogs? To what use could we ever hope to put these great deserts, or those endless mountain ranges, impenetrable, and covered to their very base with eternal snow? What can we ever hope to do with the western coast, a coast of three thousand miles, rock-bound, cheerless, uninviting, and not a harbor on it? What use have we for such a country? Mr. President, I will never vote one cent from the public treasury to place the Pacific coast one inch nearer to Boston than it now is. (Connelley 146)

Proslavery Southerners, in contrast, saw the exact opposite—the expansion of slavery as inevitable. Such ideas would lead them westward into Mexican Texas. In east Texas, they found a humid, warm climate, much like that of the US South, and cotton production—and slavery—spread rapidly across east Texas. Overwhelming the Tejano Mexican population, Anglo settlers declared Texas an independent "Lone Star Republic" in 1836. Its subsequent annexation by the United States in 1845 triggered a war with Mexico. For Mexico, the war resulted in the catastrophic loss of more than half of its national territory (Chávez).

For the victorious United States, the war seemed to be a fulfillment of everything Manifest Destiny had promised. They had defeated a weaker nation, to their eyes inherently inferior due to its Catholicism and mixed race population. Now all that formerly Mexican land lay free for the taking, and slave owners eagerly anticipated expanding their cotton empire to the Pacific. Northerners fretted about the same outcome—the spread of slavery meant the growth of its economic power and political influence. The United States had won the war, but its union was fragile, and in gobbling up more than half of Mexico, it nearly undid itself in the Civil War. Manifest Destiny had become manifest disaster, and not for the last time.

That looming sectional division rapidly became apparent in differing views of climate. A survey party, commissioned to delineate the new boundary between the two nations, offered a concise example of the sectional and climatic divide. The head of the boundary commission, New Yorker John Russell Bartlett, was sacked after committing two unpardonable crimes: compromising with his Mexican counterpart on the location of the boundary, and publicly stating that much of this new American territory appeared to be worthless. As he and his men trudged west from

El Paso through the rugged deserts of the future states of New Mexico and Arizona, his faith in the national enterprise began to falter:

> As we toiled across these sterile plains, where no tree offered its friendly shade, the sun growing fiercely, and the wind hot from the parched earth, cracking the lips and burning the eyes, the thought would keep suggesting itself. Is this the land which we have purchased, and are to survey and keep at such cost? As far as the eye can reach stretches one unbroken waste, barren, wild, and worthless. For fifty-two long miles we have traversed it without finding a drop of water. (Albuquerque Museum 37-38)

He was stripped of his command, and Congress refused to pay for the publishing of his journal, unlike the other surveys they had funded. His replacement, William Emory, while more attuned to the territorial aspirations of the Southern slaveholding interests who had pushed for war, was nonetheless likewise hard-pressed to paint an overly positive picture of the region. His best hope was that technological development—railroads—and military suppression of the Apache would render the region more amenable to development, preferably through slave agriculture (Greenberg).

Bartlett, a New Yorker, had been undermined by other members of the boundary survey, Southerners suspicious of the Yankee Bartlett. They reported his actions to Southerners in Congress, and thus forced his removal. This action was indicative of far deeper sectional divisions—differences certainly based in political, economic, racial, and social outlooks, but also no less in climatic perceptions. Slave-owning Southerners hoped to expand their lucrative slave empire into the newly-annexed Southwest, perhaps even all the way to the Pacific. They were eager to imagine that the soils and climate that had proven so salubrious in the Southeast would be found further west as well, and were in no mood to hear news to the contrary. In 1845, South Carolinian John C. Calhoun, who had served as vice president and senator, asserted that the states of the South and West occupied a single physiographic region, extending from the Atlantic coast to the Gulf coast, and from the Mississippi River Valley to the valley of the Rio Grande. The arid landscape of west Texas or New Mexico—at that point, still Mexican Nuevo Mexico—was transmogrified in Calhoun's wishful thinking into a place indistinguishable from that of humid and wet South Carolina (Smith 148). False climatic assumptions like these would help lead the United States into war with Mexico. Clearly, false delusions could have real consequences (Smith 145-54). For that matter, such delusions were born of spectacular success. In the period from 1815 to 1860, the population of the US "West"—which is to say the western Midwest, the western South, and Texas, grew from a population of one to fifteen million. In the face of such prodigious growth, more of the same seemed plausible (Belich 223).

In opposition to this myth of a Southern pastoral garden in the West was another idea, more grounded in fact, even if still a cultural construction in its own way. This was the belief that some portion of the West—perhaps just a narrow strip of land running north and south on the High Plains just east of the Front Range of the Rocky Mountains, or perhaps a much larger area, encompassing much of the West, was a "Great American Desert." Zebulon Pike's journal of his 1810 expedition cast the plains, treeless and windswept, as a desert as forbidding as the Sahara. The Stephen F. Long expedition reconfirmed this when its report was published in 1823 (Smith 175-76).

Aridity seemed to preclude the possibility of farming—a dire outcome for a nation where the idealized yeoman farm was sacrosanct. The claim that some significant part of the West was what Washington Irving termed the "Great American Desert" in his book *Astoria* (1835), an account of the western fur trade, seemed to throw the entire enterprise of western territorial expansion into doubt. His description certainly struck at any idea of a western "garden," ripe for easy settlement:

> It is a land where no man permanently abides; for, in certain seasons of the year there is no food either for the hunter or his steed. The herbage is parched and withered; the brooks and streams are dried up; the buffalo, the elk, and the deer have wandered to distant parts, keeping within the verge of expiring verdure, and leaving behind them a vast uninhabited solitude, seamed by ravines, the beds of former torrents, but now serving only to tantalize and increase the thirst of the traveler. (167-68)

Irving's book popularized the term, and the "Great American Desert" would appear on maps of the West for decades to come as a place unfit for farming and settlement. Yet Irving, like so many of his era, made more than agricultural assumptions based on aridity. A region unsuitable for farming was unsuitable for civilization, and, by inference, civilized people. If anyone did ever live in this forbidding landscape, Irving feared that the Great American Desert might serve as the problematic birthplace of "new and mongrel races, like new formations in geology, the amalgamation of the 'debris' and 'abrasions' of former races, civilized and savage; the remains of broken and almost extinguished tribes; the descendants of wandering hunters and trappers; of fugitives from the Spanish and American frontiers; of adventurers and desperadoes of every class and country yearly ejected from the bosom of society into the wilderness" (168).

In Irving's view, the Great American Desert was more than a disappointment to prospective settlers. It was a permanent impediment to settlement, to national ambitions of continental supremacy, and might in the future even serve as a homeland for warlike nomads born of a hybrid English, Spanish, and Indian ancestry who would prey on more settled and "civilized" peoples. This literary impression was followed up by many eyewitness accounts, and ultimately buttressed by government-funded science. Federal surveys, like those led by John Charles Frémont, led to the publication of survey reports full of detail, multivolume books lavishly illustrated with landscapes, flora, fauna, and maps.

In this era, maps were more than simple visual expressions of geographic information. In the early republic of the United States, they represented national ambition and the confluence of science and art. Geographic literacy—the ability to read and create maps, and the money to collect them—was proof of social and economic status. Several of the nation's founders, including George Washington and Thomas Jefferson, took pride in their experience and abilities as land surveyors. Not for nothing were so many Americans—the rich and the striving alike—pictured with maps and globes in individual and family portraits. These maps were visual manifestations of individual and collective ambitions. Maps, in short, mattered, and citizens took them seriously (Brückner).

These published reports made a significant impression on eastern readers. Historians today often see them as landmarks in the scientific exploration of western North America, but do not always realize that to many readers of the time they were more like glossy real estate brochures. The reports, in fact, often cost more to print than the surveys themselves cost to complete. That does not mean that these reports were necessarily false. Instead, it meant that many readers were only too eager for them to confirm their most optimistic imaginings about the West (Goetzmann).

Utah's Desert and the Question of Western Climate

Few readers were more interested than members of the Church of Jesus Christ of Latter-day Saints, known to most outsiders as Mormons, a new religious group which had settled in the Midwest but attracted hostility due to its unorthodox theological beliefs and social practices, including polygamous marriages between one husband and multiple wives. Looking for an escape, church leaders pored over Frémont's reports, considering locations in Texas, California, and Vancouver Island, but rejecting all of them as *too* attractive—they would soon be inundated with settlers, and the Mormons would be hopelessly outnumbered. Instead, they began to focus on a series of valleys in the eastern Great Basin, hoping that somewhere in that vast, unsettled space they could find a home that would sustain them, but not attract too much attention from others (Farmer 39-40). It was, they imagined, a place where "good living will require hard labor," and they were right (Arrington 41). However, snowmelt fed streams and rivers flowed through the region, most converging on the Great Salt Lake, a salty sea with no outlet to the ocean. That fresh water, channeled into communal irrigation projects that watered the good soils of the Salt Lake Valley and surrounding valleys, led to a successful settlement, the first Anglo-American settlement to succeed and endure in this harsh region. Native Americans, such as the Pueblo Indians, and Mexican settlers in New Spain and the Mexican Southwest, had long practiced irrigation, but for Anglo Americans it was largely new. The new Mormon Zion was certainly not an easy place, but it had its advantages in addition to the reliable streams. There was ample salt on the shores of the Great Salt Lake, hot and cold springs for drinking, bathing, and "taking the waters" for health, and the high elevation, aridity, and cold winters kept diseases such as malaria at bay. It was, in short, far healthier than their former settlements in the Midwest, which had been plagued by malaria and other illnesses. Robert Bliss, who arrived several months after the first wave of settlers, reported that "The atmosphere is pure & there has been no sickness as yet among us to speak of.... All are pleased with the climate" (Farmer 42-47, 45; Worster 1986: 74-83).

The Mormons, like other European settlers, saw "climate" as synonymous with health, or its absence. They believed that illness was spread by bad air, "miasmas" of fog, mist, or stale atmosphere. Tuberculosis and other respiratory illnesses were most often blamed on this "bad" air, but other illnesses were as well. Swamps or other places of stagnant water were viewed as public health dangers in need of draining, just as a sick person needed bleeding. While their knowledge of pathogens, bacteria, or viruses was primitive or nonexistent, their concerns were well-placed—swampy water bred mosquitoes, and mosquitoes were most definitely vectors of contagion. Settlers saw clear connections between their bodies and the surrounding environment, a link that later generations would lose, and are only now reconnecting, as twenty-first century humans increasingly worry about contaminants, from heavy metals to artificial hormones, accumulating in ecosystems and in human bodies (Valenčius).

The Mormons arrived in Salt Lake Valley in July 1847, and the Gold Rush and the conclusion of the US-Mexico War meant that the United States arrived soon after. They accordingly began petitioning to be admitted to the union as a territory and state, a goal they would not finally achieve until 1896, after formally renouncing their practice of polygamy. Polygamy, as well as the vast territory they tried to claim—essentially all of the Great Basin plus Southern California—meant that their proposal elicited little support in Washington. In fact, it was rejected for yet another reason as well—its proposed name, Deseret. According to Mormons, this was an ancient Egyptian word for the honeybee, and connoted the industry and organization which they hoped characterized their society. To eastern ears, however, that supposed meaning was trumped by a phonetic problem. "Deseret" sounded too much like "desert." The new dominions of the United States might well be arid, but calling a vast swath of them a desert was too objectionable. Though the name lingered in local usage—the church-owned newspaper is still called *The Deseret*

News today—it was jettisoned in favor of another name, Utah, derived from the Ute Indians who populated the Salt Lake Valley and much of the future state in 1847.

Combating Climate with Irrigation and Reclamation

The federal government might have vetoed Deseret, but the desert regions of the West were real enough. In response to arid climate conditions, irrigation seemed the only feasible solution. The success of the Mormon settlement was held up as proof that cooperative irrigation could be successful. While a growing chorus supported private and public irrigation schemes, there were significant divisions within the irrigation movement, or what would come to be called "Reclamation." Two individuals personified this divide: John Wesley Powell and William Ellsworth Smythe.

Powell had made a national name for himself by leading an expedition down the Colorado River, surviving the terrifying rapids in the Grand Canyon and other gorges, and surveying geography utterly unknown to whites. Powell later served as director of the United States Geological Survey. His reports combined Victorian travelogues and artful descriptions of scenery with a clear-eyed and decidedly unromantic view of the limited potential of arid western lands for settlement and agriculture (Padget). The culmination of his writings was his *Report on the Lands of the Arid Region of the United States* (1879). Powell argued that the arid West called for a new government approach, and a new kind of homesteading. Unlike in the East, homesteading land allocations would have to be much larger, and the government would have to plan large-scale irrigation projects: "To a great extent, the redemption of all these lands will require extensive and comprehensive plans ... individual farmers, being poor men, cannot undertake the task" (viii). Powell thought that the arid West, properly settled through careful planning, could be home to a small population of farmers and ranchers.

This modest, slow, and carefully planned future did not appeal to some Americans, who thought that irrigation could be transformative—it could utterly remake the West into a place as populous and agriculturally productive as the East. A vociferous champion of this view was William E. Smythe. Smythe agreed with Powell on the importance of irrigation. He, however, saw irrigation as a *deus ex machina*, a wondrous solution that could transform the desert into a garden, make farms productive, and create cities as large as any in the East. His magnum opus— printed as a book and in serial or excerpted form in numerous regional and national magazines, was *The Conquest of Arid America* (1900). In it, he essentially resuscitated Manifest Destiny for a new century. Instead of divine providence, irrigation technology was now the force that would lead Americans to their continent-conquering destiny. For Smythe, the Mormon settlement in Utah had created one of the "real utopias of the arid West" (49). He likewise praised the growth of Southern California, which had transformed from a rural backwater in the 1870s to a booming urban and agricultural region by the early twentieth century. He toured the country extolling the virtues of irrigation, telling fervent believers that irrigation had freed them from climate. He even claimed that aridity was a virtue, because it forced individualistic Americans to work together. Better yet, aridity "compels the use of irrigation. And irrigation is a miracle!" (40) Irrigation would create wealth, encourage civilization and science, replace isolated farms with irrigated cities and farmlands interspersed together, merging city and country into an irrigated utopia.

This was heady stuff indeed, and readers eagerly believed what Smythe was promising. So did the nation's political class. In 1902, Congress passed the Newlands Reclamation Act, authorizing the federal funding of dam and irrigation projects. Smythe crowed that the Reclamation Act "is perhaps the only measure in the history of American legislation enjoying a popularity so absolute and unquestionable that the only possible controversy between the two parties was as to which was entitled to the greater credit for bringing it to pass" (287). The first project completed under the new law was the Roosevelt Dam on the Salt River in Arizona. Anglo settlers renamed the Salt River Valley the Valley of the Sun, planted lawns and eastern greenery to offset the stark landscape of the Sonoran Desert, and metropolitan Phoenix was born. By 2010, Phoenix—located in the middle of one of the driest deserts on earth—had surpassed Philadelphia to become the fifth largest city in the United States. Powell likely would have found this unsustainable and foolhardy, but his vision of the West had lost, and Smythe's had won (Logan).

Yet Powell and Smythe, despite their divergent views, both accepted that the West was indeed fundamentally arid. Their divergence concerned the degree to which irrigation could alter this condition. But what if this fact could be occluded, or even undone? Could aridity be erased? Was the aridity of the region overstated? Some writers tried to assert that aridity was a mirage. Others would assert that it could be eliminated altogether through human action. Belief in their claims would prove one of the greatest—and most foolhardy—acts of climatic delusion in US history.

William Gilpin and the Erasure of the Great American Desert

The person most responsible for erasing the Great American Desert was William Gilpin. He was both an individual and a representative of an entire class—the western booster who sold the lands of the West regardless of reality. Gilpin should have known better—he had traveled to the far West with John C. Frémont's 1843 expedition, travelling all the way to Fort Vancouver. He fought in the US-Mexico War and in campaigns against the Comanches and Pawnees. During the Civil War, he served as the first territorial governor of Colorado. All that lived experience, however, would prove no impediment to his assertions about the far West. Wallace Stegner, a twentieth-century author and environmentalist, asserted that Gilpin "looked clear over the continent of facts and into prophecy." According to Stegner, Gilpin "saw the West through a blaze of mystical fervor, as part of a grand geopolitical design, the overture to global harmony; and his conception of its resources and its future as a home for millions was as grandiose as his rhetoric, as unlimited as his faith, as splendid as his capacity for inaccuracy" (2). For Gilpin, it was all simply a matter of perspective—the right point of view. He presented himself as a sensible, educated, and experienced man, one who had coolly weighed all of the data and evidence available for the landscapes, climate, and agricultural potentialities of the West, and found a revelation: "To the American who assembles within his mental glance every detail of our entire country, from a position correctly selected and rightly understood, a vision of unparalleled splendor is unveiled" (91).

According to Gilpin, everything prospective settlers thought they knew about the Great American Desert was wrong. Yet Gilpin did not begin by simply weaving myth out of whole cloth. His assertions did usually have some marginal basis in fact, but his assertions would stray ever further from that fact until his claims bordered on the nonsensical. Gilpin proclaimed that "I am struggling to narrate faithfully the homespun facts of nature: to exaggerate is far from my intention" (17-18). He did not claim that the West was not arid—he claimed instead that irrigation and cattle-grazing would triumph over that aridity, and even that the region's aridity, instead of a hardship, would actually save labor and make money for farmers.

Gilpin merged avarice, wishful thinking, "science," religion, and Manifest Destiny into a heady mix. For Gilpin, the West was first and foremost about the future, and that future was full of vast, even awesome, promise. His expansive geography also matched the acquisitive worldview of Anglo-Americans, interested in spreading their national domain as far as possible. According to Gilpin, the "Plateau of the Table Lands" between the Rockies and Sierras included not merely the Great Basin, but extended south to the Valley of Mexico and all the way to the Andes. This territory, he asserted, "appears to me the most interesting, the most crowded with various and attractive features, and the most certainly destined eventually to contain the most enlightened and powerful empire of the world" (103).

Gilpin asserted that the Great American Desert was a mirage he had come to dispel forever:

The scientific writers of our country adhere with unanimity to the dogmatic location somewhere of "*a great North American desert*." Travellers, under their promptings, especially search for it. It has been located *seriatim* in advance of the settlements, in Kentucky, in the Northwest, in Missouri, upon the Plains, in California. NO explorer or witness who has failed to find a desert is allowed credence or fame. Yet there is none, either in North or South America; nor is the existence of one possible. This dogmatic *mirage* has recently receded from the basin of the Salt Lake; it is about to be expelled from its last resting place, the basin of the Colorado. (49)

In its place, Gilpin presented a pastoral paradise, an escape from labor, from the pressures of the market, and, ultimately, from reality itself. Instead of an arid grass-land, where vegetation dried out under the summer sun, "these delicate grasses grow, seed to the root, and *are cured into hay upon the ground* by the gradually re-turning drouth" (72). The dry climate, it seemed, helpfully eliminated the need for harvesting or baling hay. The vast herds of bison and other grazing animals, and the Indian subsistence they supported, were ample proof, in Gilpin's view, that they could be seamlessly replaced with domestic cattle:

> The immense population of nomadic Indians, lately a million in number, have, from immemorial antiquity, subsisted exclusively upon these aboriginal herds.... From this source the Indian draws exclusively his food, his lodge, his fuel, harness, clothing, bed, his ornaments, weapons, and utensils. *Here is his sole dependence from the beginning to the end of his existence.* (72-73)

For farmers struggling to make payments on land, or sell crops at a profit, this vi-sion of a climate which offered financial independence must have seemed appeal-ing indeed. He also offered assurances that this would be no rude wilderness, but one of the civilized places of the earth:

> The climate of the *Great Plains* is favorable to health, longevity, intellectual and physical development, and stimulative of an exalted tone of social civiliza-tion and refinement. (73)

Gilpin erased the reality of the high plains with a fictional landscape of wish ful-fillment, erasing all of the travails ordinary Americans faced in the post-Civil War era. Good land in the East was taken, the Civil War had wreaked havoc on the economy of the South, and the nation was enduring a series of economic crises that lasted from the 1870s through the 1890s. Here, it seemed, was a solution—the West as an escape to a better life. Later "boosters" of climate and region—of Southern California and the Southwest in the 1880s to 1920s, for example, who sold an agricultural paradise, homebuyers' Eden, and health refuge, owed some-thing to Gilpin. They aimed more at vacationers and homebuyers than homesteading farmers, but their consumerist, boosterist climate efforts began here—subsequent booster efforts were merely extensions, intensifications, and elaborations (Wrobel).

Likewise, his own efforts could be read back into the past. If Gilpin served as a premonition of climate-related real estate booms, his writings also preserved the religious, political, and ideological underpinnings of Manifest Destiny by transforming them into "scientific" language and assertions. Gilpin became rapturous when contemplating the "untransacted destiny" of the American people:

> In the current of ages, mysteries become *sciences*. Vague speculation, long fermenting, and perplexed by obscure doubts, produces facts. These crystallize into precious truth…. *The American people* now reach and cross the threshold, where they emerge from the twilight [and] enter into the full and perpetual light and promise of *political and social science*. (124, 99)

This "science" of climate, which Gilpin ascribed to both terrain and global temperature gradients, he also applied to assertions about race and civilization. According to Gilpin, the world's great civilizations developed in "succession along the undulating zone of the northern hemisphere of the globe, within the isothermal belt. They form within it a continuous zodiac from east to west. These *empires* are the Chinese, the Indian, the Persian, the Grecian, the Roman, the Spanish, the British, finally, the *republican empire* of the people of North America" (106). Civilization had been born of climate, and its final, perfected product was the United States. For Gilpin, America's global destiny would be fulfilled by the construction of a transcontinental railway, in turn only a part of a globe-encircling "Cosmopolitan Railway," which would unite all the world's continents, commerce, and politics, and make the United States the hub and center of the world. His book even contained maps of this fanciful railway, stretching across the continents and linking them all to North America.

From Erasing Climatic Realities to Imagining Climates Remade through Human Action

Gilpin sold climate—putting the best face on a problematic region. Soon, however, an even more radical mindset gained currency. What if the innate climatic characteristics of a region could be changed? Was such a thing even possible? If so, how? Settlers had speculated for some time that perhaps climate was changeable—not only that weather patterns might change naturally over time, but that human actions could alter the climate. Even before the US-Mexico War, Josiah Gregg, who had traded on the Santa Fe Trail, speculated about this prospect. He asserted that fire, rather than aridity, had discouraged the growth of trees on the plains. With those fires suppressed, "we are now witnessing the encroachment of the timber on the prairies" (362). Yet he went much further than this speculation about past vegetation:

> The high plains seem too dry and lifeless to produce timber; yet might not the vicissitudes of nature operate a change likewise upon the seasons? Why may we not suppose that the genial influences of civilization—that extensive cultivation of the earth—might contribute to the multiplication of showers, as it certainly does of fountains? Or that the shady groves, as they advance on the prairies, may have some effect on the seasons? At least, many old settlers maintain that the droughts are becoming less oppressive in the West. The people of New Mexico also assure us that the rains have much increased of latter years, a phenomenon which the vulgar superstitiously attribute to the arrival of the Missouri traders. Then may we not hope that these sterile regions might yet be thus revived and fertilized, and their surface covered one day by flourishing settlements to the Rocky Mountains? (362)

Humans had in fact speculated about changing climate since ancient times. Greek philosopher Theophrastus, born on Lesbos around 372 BC, noted places where humans had changed the local landscape by draining wetlands or rerouting rivers, and the result had been a marked warming or cooling in the local vicinity. In one locality, Larissa, in Thessaly, for example, the draining of a large expanse of wetlands seemed to result in a much cooler climate regime, too cold, in fact, for the olive trees which had once been a hallmark of the area. Philosopher Albert the Great, born around 1200 in Lauingen an der Donau, in Bavaria, thought that cutting down trees could change the climate. English author John Evelyn, author of *Silva: or, a Discourse of Forest-Trees* (1664) noted that climates could be changed by deforestation. He believed that unregulated forest growth led to excess rain and mist, and unhealthy places that needed to be cleared. Once made ready for farming and grazing, "those gloomy tracts are now become healthy and habitable" (Glacken 129-30, 316, 485-88).

On a bigger geographic scale, colonists in North America noted that their deforestation of large areas seemed to create warmer temperatures, and scientists in both North America and Europe remarked at length about this phenomenon, with most taking it as fact. Others claimed that the clearing of forests seemed to reduce weather extremes, or even the intensity of storms on land and at sea. If clearing forests seemed to warm climates, some suggested that planting them might have the opposite effect. This led to speculation that large-scale tree-planting might effect a salubrious change in desert climates (Glacken 542, 659-70, 669).

Critics of the idea of climate changed by human action fell back on their belief in a divine design within nature. If God had in fact made it, He had made it to His liking, and humans were presumptuous to think they could alter it. Yet belief in a human role in changing climates—and seeming proof of them—was common in North America in the eighteenth and early nineteenth centuries. Mills in Vermont that had been productive were abandoned when once-reliable streams could no longer supply sufficient water power. Crop-producing farmlands where swamps had once been seemed proof that rainfall patterns had changed—perhaps air and water temperatures had as well (Glacken 689-90). French writer Constantine François Chasseboeuf Volney took as "incontestable fact" the assertions from a variety of American writers—including Thomas Jefferson—that the new nation's climate seemed to have changed in response to human actions:

On the Ohio, at Gallipolis, at Washington in Kentucky, at Frankfort, at Lexington, at Cincinnati, at Louisville, at Niagara, at Albany, everywhere the same circumstances have been repeated to me: 'longer summers, later autumns, and also later harvests; shorter winters, snows less deep, and of shorter duration, but cold not less intense.' And in all the new settlements these changes have been represented to me not as gradual and progressive, but as rapid and almost sudden, in proportion to the extent to which the land is cleared. (Glacken 690)

Of course, all of this reportage of a climate changing rapidly due to human actions ignored the fact that these landscapes were hardly "virgin land" only recently subject to human actions. European settlers' awareness of Indian agency usually varied in direct relation to their greed. Since "vacant" unimproved land—any tract without "improvements," such as fences, plowed croplands, or houses—was considered free for the taking under English Common Law, many settlers proved adept at ignoring landscapes shaped by Indian actions. The terrible toll of European diseases erased Native American populations with terrifying rapidity, and this absence made European ignorance even more prevalent. The same was unfortunately long true for many historians as well. More recent historical studies, aided by environmental history, archaeology, anthropology, ecology, and other disciplines, have painted a very different picture of the eastern woodlands and Atlantic seaboard of North America, one in which Indians maintained open grazing lands and thinned forests with fire, planted crops, and encouraged the growth of nut-bearing trees. The breathless accounts of natural abundance reported by early European settlers— the teeming game, abundant fish and shellfish, and forests filled with chestnut trees and seemingly innumerable passenger pigeons and other birds—were all the products not of an uninhabited, Edenic continent, but rather a continent long tended for human benefit, but then suddenly depopulated. Recent studies indicating that precontact Indians in eastern North America were taller, healthier, and had better diets than contemporary Europeans strongly suggest that this natural abundance was real, and directly shaped by centuries of Indian manipulation of the landscape. European settlers might argue whether or not they were shaping the landscape and changing the climate. The idea that native peoples actively shaped the environment is, however, no longer a matter of debate, but established fact (Cronon; Mann).

The Great Salt Lake and Scientific "Proof" that Rain Follows the Plow

As settlement and exploration moved westward, Euro-Americans continued to speculate about changes in climate, whether in the distant past, or more recent and potentially due to human causality. Just as they had long read unfamiliar landscapes to make assumptions about climate, they now read changes in landscapes—changes in vegetation, the amount of water in lakes and streams, and other natural characteristics. One landmark that drew an exceptional amount of attention was the Great Salt Lake of Utah. Mormon settlers found it useful as an endless source of salt, and it eventually became a destination for local recreationalists and visiting tourists who enjoyed the novelty of floating in the salty water, so saline that human bodies could not sink in it, but instead bobbed atop the water like corkwood. Scientists and geologists were attracted to the lake as well, albeit for different reasons. On the mountains surrounding the lake, the islands within it, and even on mountain ranges in distant parts of the Great Basin, there were terraces—ancient shorelines—as clear as rings in a bathtub, and some many hundreds of feet above the current level of the lake. It soon became evident that the Great Salt Lake was only a small remnant of a huge inland sea, one that had apparently waxed and waned for eons before any European had ever seen it.

In 1849, Howard Stansbury, of the Army Corps of Topographical Engineers, made note of these ancient, elevated shorelines, which locals began calling "benches." More surveys and scientists followed—Clarence King's Fortieth Parallel Survey, F.V. Hayden's Geological and Geographical Survey of the Territories, John Wesley Powell and the Powell Survey, and George M. Wheeler's Geographical Survey West of the One Hundredth Meridian. Various theories were put forward to explain the lake's fluctuating volume. Perhaps the mountains surrounding the lake had risen, taking old shorelines upward with them. Perhaps instead the ground under the lake itself had risen, displacing its waters. Perhaps the lake had somehow formed an outlet with the ocean via a now-vanished river or a hidden, subterranean outlet. The first whites to sail boats across the lake had heard unsettling reports of a whirlpool somewhere in it that siphoned the lake's waters away to the sea—and might carry any unfortunate sailors with it as well. A more mundane explanation, however, soon took precedence—at some point in the past, the climate had become more arid, and as a result the lake had gradually dried up and shrunken to its current form.

This was undoubtedly a subject of interest for scientists, but Mormon settlers in the area found it a more pressing issue. Could the lake dry up completely, and the Mormon Zion become a barren desert? Conversely, could the lake's waters rise once more, and inundate Salt Lake City? Suddenly, the lake's water level was of a great interest to a great many people. In the absence of precise measurements, settler accounts were helpful. Locals recalled years when it had been easy to ride a horse or herd cattle across shallow expanses of water to reach islands near the lake's shores that had good grazing land. From 1865, however, these journeys could sometimes only be made by boat. Such accounts, paired with more precise measurements, painted a potentially unsettling picture. From 1847, when the Mormons arrived, until 1865, the lake had fluctuated within a narrow range. By the early 1870s, however, the lake was clearly growing. Its surface area was perhaps as much as 20 percent larger than the lake Stansbury had first surveyed in 1849.

This seemed to provide scientific credence for a widespread belief among the Mormons that wherever they settled, an increasing water supply followed. They might ascribe this to providence, but John Wesley Powell suspected a more mundane answer. The Mormon settlers, like other Anglo American settlers, had cut down trees, and their sheep and cattle had eaten native grasses. They had straightened streams, and drained wetlands. All this, Powell suspected, meant that more runoff from rain and snowmelt was reaching the lake. That, plus perhaps a periodic uptick in precipitation, explained the rising lake waters (Morgan 305-14).

That supposition, however, soon lost public currency to a theory more audacious, and more attuned to Anglo-American triumphalism and Manifest Destiny. This idea posited that the Mormon settlers, reviled as they might have been by most other American Protestants, were correct. Increased precipitation followed settlement. Or, in words that would come to haunt the environmental history of the United States, "rain will follow the plow." In fact, this theory received the qualified support of a scientist in the Hayden survey party, Cyrus Thomas.

Thomas noted the longstanding climatic conundrum of aridity—the basic problem that confronted would-be settlers of the West: "The troublesome factor in the great problem of the development of the agricultural capacity of the vast western plains is the supply of water.... Furnish this, and the 'Great American Desert' of old geographers will soon become one mighty field of flowing grain" (Hayden 140). Thomas reported that he had encountered what others had also reported—that new areas of settlement seemed to soon discover that water supplies were increasing. He at first discounted these reports, but paid more attention when he heard the same stories repeated in each new area he visited:

Streams bearing down heavier volumes of water than formerly; others becoming constant runners which were formerly in the habit of drying up; springs bursting out at points where formerly there were none; acequias [Mexican irrigation ditches] allowed to go to decay because they have not been needed, &c. Even the Arkansas [River], as late as 1862 and 1863, was dry, from Pawnee to the Cimarron crossing, but such a thing has not been known since.... [It] is the uniform assertion of those who have resided in the Territory for ten or twelve years or more, that for six or seven years past there has been a gradual increase of rain. It is a common expression of the Mexicans and Indians that the Americans bring rain with them. (Hayden 140-41)

These repeated accounts led him to a conclusion with profound consequences:

All this, it seems to me, must lead to the conclusion that since the Territory has begun to be settled, towns and cities built up, farms cultivated, mines opened, and roads made and traveled, there has been a gradual increase of moisture.... I therefore give it as my firm conviction that this increase is of a permanent nature, and not periodical, and that it has commenced within eight years past, and that it is in some way connected with the settlement of the country; and that, as the population increases, the amount of moisture will increase. (141)

Thomas went on to qualify this statement. The increased water supply might be the result of a temporary increase in rainfall. It could be a product of some localized peculiarity of the places he had studied, and might not be replicated elsewhere. He readily admitted that the small number of settlers, in comparison to the vastness of the territory in which that had settled, made a human role in climate change seem unlikely. Moreover, he called for continued and increased study to verify if his supposition was correct:

Such a theory may, and doubtless will by some, be considered chimerical, but before it is condemned some effort to confirm or refute the testimony given ought to be made.... Even should a more thorough examination overturn and reverse the testimony I have adduced, his labor need not be lost, as he could, while proceeding with this, gather a host of facts in regard to the agricultural capacity of our Territories, which would be of great value to the stream of emigrants pressing westward from the States. (141-42)

Thomas the scientist had stated that there might be causality in the relationship between settlement and rainfall. He had called for further study to empirically prove or disprove his assertions. Yet the boosters and all the settlers and investors who wanted to believe in them paid no heed to those cautious qualifications. All they heard was the siren call that "rain follows the plow." For neither the first nor last time, journalists and politicians freely appropriated and manipulated science for their own purposes. Charles Dana Wilber, a journalist and land speculator who had founded the settlement of Wilber, Nebraska in 1873, seized on Thomas' words, eliminating any scientific caution and trumpeting only the incredible news, now verified though science, that "rain follows the plow." Deserts, according to Wilber, existed at all only because humans were passive: "By the repeated processes of sowing and planting with diligence the desert line is driven back, not only in Africa and Arabia, but in all regions where man has been aggressive, so that in reality there is no desert anywhere except by man's permission or neglect" (69). He asserted that formerly verdant arable lands, such as the valley of the Euphrates in Mesopotamia, had been "lost" to desert due to human laziness, or the decline of civilizations. Other possible causes—overgrazing, soil erosion, or changes in climate, were dismissed. Anglo Americans, armed with their own labor, the laws of science, and the benevolence of God, had the power to transform any landscape:

> It requires only the condensing surface of growing verdure; it may be of trees or shrubs or growing grain, over large areas; or, in short, just such a changed surface as man necessarily brings about as a tiller of the soil, to compel the moisture to take cloud forms in the atmosphere, instead of being dispersed by the daily radiation of solar heat. Everywhere, under these new conditions of husbandry, the clouds will gather into larger clouds, and overspread the heavens; and the impending shower will fall upon the farm and garden, not by a grace or fortuity, but by an eternal law. Yet, in this miracle of progress, the plow was the avant courier—the unerring prophet—the procuring cause. Not by any magic or enchantment, nor by incantations or offerings, but, instead, in the sweat of his face, toiling with his hands, man can persuade the heavens to yield their treasures of dew and rain upon the land he has chosen for a dwelling place. (70)

Wilber blithely asserted that deserts and any other arid regions were merely "temporary conditions of the earth's surface," which could "through the industry as skill of man, be changed into fertile and productive fields" (71). While praising the benevolence of God and nature, Wilber reserved special disdain for anyone who dared disagree with his claims:

> With the power in our own hands to make the wilderness and waste places glad, and to make even a desert blossom as a garden with roses, I cannot find suitable words of censure for those who falsely represent our great national domain as being in most respects not only arid and useless for the abode of man, but by a

physical necessity forever forced to be under desert; conditions that imply only the sustenance and support of a few nomadic herdsmen. (71)

From Manifest Destiny to Manifest Disaster

Wilber had mined the words of scientists to make his claims. Now, however, he castigated anyone who might suggest that his assertions warranted some scientific caution. His book, and others likewise claiming that "rain follows the plow" drew large numbers of settlers and speculators onto the plains—and towards disaster. A speculative boom and bust in cattle—in which millions of dollars were lost and hundreds of thousands of cattle died—occurred on the High Plains in 1886, the result of an especially harsh winter and overly-optimistic assumptions about how many cattle the native grasses of the plains could support. Aided by global investors and newly-invented barbed wire, ranchers filled the prairie with cattle on the assumption that they could graze through the winter. The cattle instead froze and starved en masse, piling in ghoulish mounds alongside the barbed wire fences. The calamity triggered a depression among plains ranchers that lasted twenty years (Nugent 75).

This calamity scared away eastern and international investors for a time, but not settlers. From the passage of the Homestead Act, from 1862 to 1900, 1,400,000 people applied for homesteads in the West. From 1900 to 1913, another million would do so, more than twice as many per year as before 1900. With the best land already taken, these later settlers moved into arid regions, areas of high elevation, and places where rain and favorable temperatures were the least reliable. Many of them started a new life right in the heart of the lands that had once borne the foreboding title of the "Great American Desert." In the first two decades of the twentieth century, nearly three million people moved into Texas and Oklahoma. More than a million settled in the states of Kansas, Nebraska, and North and South Dakota in the same years. Wheat farms, grain elevators, farm houses, and windmills pumping groundwater for surface irrigation rose where prairie grasses had once grown. Many of these were built by eastern urbanites and European immigrants; Germans, Scandinavians, East Europeans, and others who had been drawn by the promise of American plenty. Few of them had any experience with arid farming, and fewer still knew the dire environmental dangers of the region they had moved to (Nugent 131-41).

They would be struck by a combination of economic and environmental disasters. Technology—first water-pumping windmills, then tractors, harvesters, and other mechanical innovations, allowed agricultural productivity to grow. Production increased further due to increased global demand during World War I. After the war, however, demand plummeted. Prices for staple crops such as corn and wheat collapsed, and farmers found themselves with crops worth less than it would cost to harvest and ship them to market. Even before the market collapsed, farmers across the region were encountering trouble, from windmills that no longer brought up groundwater, to soil that lost nutrients due to over-farming. The settlers, or in some cases their children or even grandchildren, began to flee. Some counties and towns on the High Plains in both the US and Canada had lost a third of their population by 1925. One woman, a child during this exodus, recalled that "nature grew tired of her benevolence," and as a result "the covered wagons rolled again— eastward" (Nugent 194). This began a migration that on some parts of the plains has never stopped. In central Nebraska, for example, the population at the end of the twentieth century was less than half what it had been in 1930 (Nugent 188-95).

And something far worse still loomed in the future. Environmental history unfortunately often seems to play a limited, largely invisible role in the popular historical narrative of the United States. Such is not the case with the Dust Bowl. Overgrazed and over-plowed soil, no longer rooted by native grasses, dried out in a prolonged drought in the early 1930s. In 1933, and for several years thereafter, windstorms made the soil of the southern Great Plains take flight. It rained down on Chicago and Washington, DC, and even on ships in the Atlantic. On the plains, anyone caught outside in a dust storm could be blinded and choked to death. Historians and scientists still argue about the combination of environmental and economic forces that conspired to create the Dust Bowl. For the people fleeing its effects, the cause was a moot point. They fled for their lives, or at least for their economic survival. They joined the plains exodus already underway, and actually totaled a smaller number than those who left during the 1920s, though their plight drew much more media attention, particularly the "Okies" who headed west to California like the family in John Steinbeck's novel, *The Grapes of Wrath* (1939). Like the exiles of the 1920s, they fled farms to take wage work in cities or in large agribusiness enterprises instead. The homesteading dream that had motivated American settlers for more than a century was dead. Yet the Dust Bowl, for all its human tragedy and ecological destruction, was only the final act in a much longer continental drama, one that stretched back to before the Homesteading Act, or the influx of settlers in the late nineteenth century (Worster 1982; Gregory).

Speaking at a decidedly more triumphal moment in US history, at the Columbian Exposition of 1893 in Chicago, which was built to commemorate the four hundredth anniversary of Columbus' arrival in the Americas in 1492, historian Frederick Jackson Turner gave an address on "The Significance of the Frontier in American History." In it, he noted that the US Bureau of the Census had declared the US frontier closed in 1890—undoubtedly surprising news to all those homesteaders who arrived after that date. His famous "frontier thesis" was fundamentally about culture, society, and politics, but not the natural environment. To the degree that he considered the frontier to be an active environmental factor in history, it was one which existed primarily as a foil for human efforts; a test to be passed and overcome, rather than something that had actively shaped the society of the United States. We can instead see that the environment—or in this particular case, climate, and perceptions and fantasies of climate—played a central and active role in US history. Similar stories can be told elsewhere, from arid Australia to the Russian Steppe. If, to borrow from Turner, we reflect upon the significance of climate in history, we can see how it shaped development, settlement, the ambitions of individuals, regions, and nations, political debates, environmental knowledge, and environmental perceptions. No less importantly, we can see how economics, politics, religion, and science all shaped how humans—from individuals to entire nations—perceived and imagined climate. If we consider how, in one nation, a confluence of settler observations, scientific speculation, promotional profiteering, and national wish fulfillment transmogrified the problematic ideology of Manifest Destiny into an ecological catastrophe, a disaster made manifest, we can perhaps better comprehend the climate perceptions and debates of the present, their connections to the past, and what they may portend for the global future.

Works Cited

Albuquerque Museum. *Drawing the Borderline: Artist-Explorers of the U.S. Mexico Boundary Survey*. Albuquerque: The Albuquerque Museum, 1996.

Arrington, Leonard J. *Great Basin Kingdom: An Economic History of the Latter-day Saints, 1830-1900*. Cambridge: Harvard University Press, 1958.

Belich, James. *Replenishing the Earth: The Settler Revolution and the Rise of the Anglo-World, 1783-1939*. Oxford: Oxford University Press, 2009.

Brückner, Martin. *The Geographic Revolution in Early America: Maps, Literacy, and National Identity*. Chapel Hill: University of North Carolina Press, 2006.

Brune, Michael. "Generation Hot." *Sierra Club Blog* 4 February 2011. <http://sierraclub.typepad.com/michaelbrune/2011/02/generation-hot-coal-epa.html/>.

Chávez, Ernesto. *The U.S. War With Mexico: A Brief History with Documents*. New York: Bedford/St. Martin's, 2007.

Connelley, William E. *Doniphan's Expedition and the Conquest of New Mexico and California*. Topeka, KS: Published by the author, 1907.

Cronon, William. *Changes in the Land: Indians, Colonists, and the Ecology of New England*. New York: Hill and Wang, 1983.

Farmer, Jared. *On Zion's Mount: Mormons, Indians, and the American Landscape*. Cambridge: Harvard University Press, 2008.

Gilpin, William. *Mission of the North American People, Geographical, Social, and Political*. Philadelphia: J.B. Lippincott and Co., 1873.

Glacken, Clarence C. *Traces on the Rhodian Shore: Nature and Culture in Western Thought from Ancient Times to the End of the Eighteenth Century.* Berkeley: University of California Press, 1967.

Goetzmann, William. *Exploration and Empire: The Explorer and the Scientist in the Winning of the American West.* New York: Knopf, 1966.

Greenberg, Amy S. "Domesticating the Border: Manifest Destiny and the 'Comforts of Life' in the U.S.-Mexico Boundary Commission and Gadsden Purchase, 1848-1854." *Land of Necessity: Consumer Culture in the United States–Mexico Borderlands.* Ed. Alexis McCrossen. Durham: Duke University Press, 2009. 83-112.

Gregg, Josiah. *Commerce of the Prairies.* Ed. Max L. Moorhead. Norman: University of Oklahoma Press, 1954.

Gregory, James. *American Exodus: The Dust Bowl Migration and Okie Culture in California.* New York: Oxford University Press, 1991.

Hayden, F.V. *Preliminary Field Report of the United States Geological Survey of Colorado and New Mexico, 1869.* Washington, DC: Government Printing Office, 1869.

Horsman, Reginald. *Race and Manifest Destiny: The Origins of American Racial Anglo-Saxonism.* Cambridge: Harvard University Press, 1981.

Irving, Washington. *Astoria, or Anecdotes of an Enterprise Beyond the Rocky Mountains.* New York: John B. Alden, Publisher, 1883.

Logan, Michael F. *Desert Cities: The Environmental History of Phoenix and Tucson.* Pittsburgh: University of Pittsburgh Press, 2006.

Mann, Charles C. *1491: New Revelations of the Americas before Columbus.* New York: Knopf, 2006.

Morgan, Dale L. *The Great Salt Lake.* Indianapolis: Bobbs-Merrill Company, 1947.

Nugent, Walter. *Into the West: The Story of Its People.* New York: Knopf, 1999.

Orsi, Jared. *Hazardous Metropolis: Flooding and Urban Ecology in Los Angeles.* Berkeley: University of California Press, 2004.

Padget, Martin. *Indian Country: Travels in the American Southwest, 1840-1935.* Albuquerque: University of New Mexico Press, 2004.

Powell, John Wesley. *Report on the Lands of the Arid Region of the United States, with a More Detailed Account of the Lands of Utah.* 2nd ed. Washington, DC: Government Printing Office, 1879.

Smith, Henry Nash. *Virgin Land: the American West as Symbol and Myth.* Cambridge: Harvard University Press, 1950.

Smythe, William E. *The Conquest of Arid America.* Revised Ed. New York: Macmillan Company, 1907.

Stegner, Wallace. *Beyond the Hundredth Meridian: John Wesley Powell and the Second Opening of the West.* Boston: Houghton Mifflin Company, 1954.

Valenčius, Conevery Bolton. *The Health of the Country: How American Settlers Understood Themselves and Their Land*. New York: Basic Books, 2002.

Wilber, Charles Dana. *The Great Valleys and Prairies of Nebraska and the NorthWest*. Omaha, NE: Daily Republican Print, 1881.

Worster, Donald. *Rivers of Empire: Water, Aridity, and the Growth of the American West*. New York: Pantheon Books, 1986.

———. *Dust Bowl: The Southern Plains in the 1930s*. New York: Oxford, 1982.

Wrobel, David M. *Promised Lands: Promotion, Memory, and the Creation of the American West*. Lawrence: University Press of Kansas, 2002.

Why is the United States a Laggard in Climate Change Policy?

Andreas Falke

Introduction

With the election of Barack Obama, hopes were high that the United States would end the obstructionist stance in international climate change policy that had characterized the George W. Bush presidency. During the campaign, candidate Obama pledged to work for a cap-and-trade system, but he deliberately left most details vague. In his first message to Congress, he urged the US legislature "to send me legislation that places a market-based cap on carbon pollution and drives the production of more renewable energy in America" (White House). The House of Representatives complied with the president's request and in June 2009 passed the American Clean Energy and Security Act (ACES), which contained at its core a complex cap-and-trade-system that was intended to reduce CO_2 emissions by 17 percent of 2005 levels by 2020. It also included extensive provisions to support renewable energy and energy efficiency, to allow for carbon offsets, and to introduce border measures to protect American energy-intensive industries from competition from countries such as China that would not undertake comparable efforts (Pew Center on Climate Change). The bill passed by a small margin.[1] It was highly complex and unwieldy, and the long term consequences in terms of its benefits and costs, and its losers and winners, were hard to gauge (Congressional Research Service 2009b). But the vote was historic in that it was the first cap-and-trade bill that had achieved a majority in a congressional floor vote.

In the Senate, climate change legislation was more problematic and various bills did not make enough progress for President Obama to arrive at the climate change summit at Copenhagen with a credible commitment to national action that could serve as a basis for international agreement. This was one of the many reasons that the Copenhagen summit yielded few tangible results. The Obama administration continued to press the Senate to come up with comparable legislation, but efforts failed due to deteriorating economic conditions, united Republican opposi-

---

[1] The margin was 219:212, with eight Republicans voting for it. The South and the Midwest voted against the bill, the South overwhelmingly so. In the Midwest a substantial number of Democrats voted against the bill (Washington Post).

tion, and waning enthusiasm among those Democrats in the Senate who hailed from coal, oil, and automobile states and feared repercussions at the polls (New York Times 2010b).

Climate change legislation had stalled. For the Obama administration, regulatory action was the only remaining option (Congressional Research Service 2009b). It allowed the Environmental Protection Agency (EPA), on the basis of the Clean Air Act, to classify CO_2 as a pollutant to be reduced, a move that was declared legal after a historic Supreme Court decision in Massachusetts vs. EPA (US Supreme Court), but which was not necessarily popular in Congress. Climate change policy had arrived at an impasse again, despite many positive signs, among them impressive activism on the state and local level, the general support of public opinion, and a shift in a number of industries and corporations towards climate change action, and of course, above all, a Democrat in the White House and Democratic majorities in both houses of Congress.

While it is hard to argue that the United States had not moved beyond the policy obstruction or negligence of the Bush period, the experience of US climate change policy during the first twenty months of the Obama administration was surely sobering. The United States is certainly more activist than under the Bush administration, as President Obama's high-level intervention at the Copenhagen climate summit in 2009 showed (Antholis et al.). But the impression of the United States as a laggard was hardly erased. Rather, the significant gains of the Republicans in the 2010 mid-term elections, giving them control of the House and reducing the Democrats' majority in the Senate, are likely to cement the outlier status of the United States. This chapter explores the reasons for the continuing laggardness of the United States. It does this by examining the development of official US government positions on climate change since the late 1980s, demonstrating that at the outset of the international debate, the United States was in the OECD mainstream in international climate change negotiations, but was then confronted with constraints that pushed it to laggard status. The paper analyzes four dimensions which can account for the laggardness: public opinion, the nature of the political process and the political system in the United States, industry influence and the rise of competitiveness concerns, and last but not least, the nature of diplomacy under the UN process as defined by the United Nations Framework Convention on Climate Change (UNFCCC).

Presidential Activism on Climate Change: Hopeful Beginnings, but Meager Results

When the climate debate first broke as a public policy issue in the 1980s, there was little indication that the United States would drag its feet on climate change. Climate change received significant presidential attention. Even the Reagan admini-stration, one of the least environmentally friendly governments, acknowledged the issue, and President Reagan signed the Global Climate Protection Act that, for the first time, required the US government to devise plans to stabilize greenhouse gas emissions (Pommerance). The Reagan administration was also instrumental in cre-ating the Intergovernmental Panel on Climate Change (IPCC) in 1988, thus helping to create the institution that would most forcefully present the scientific case for the reality of climate change, even though the Reagan administration saw studying the problem as an alternative to action (Agrawala). The first Bush administration displayed a greener attitude. During the campaign, Bush highlighted the problem of climate change and the need for action. Once elected, he appointed in William Reilly an activist administration to the EPA, gave enhanced support for climate change research, and signed the Clean Air Act, which primarily dealt with conven-tional forms of air pollution and acid rain, but would at some later date, after the intervention of the Supreme Court, serve as a basis for the regulation of greenhouse gas (ghg) emission. Bush's powerful chief of staff, John Sununu, put the brakes on further White House action as contemplated by the president and William Reilly. Sununu's move also reflected growing industry unease about tackling the problem aggressively. Only after Sununu's departure did the Bush administration play a more constructive role at the first Earth Summit in Rio de Janeiro. Although there were no hard commitments or targets except for a vague pledge to stabilize 2000 emissions at the 1990 level, the United States signed the Framework Convention, which created the international architecture for future international agreements to curb ghg emission (Antholis et al. 25-29; Weart 2009a). The most vocal critic of the Bush administration's defensive line in Rio was democratic senator Al Gore. Throughout his entire career, Gore had kept abreast of climate change issues, and he was an early advocate of government funding for research, working through hearings and other actions to make climate change a public policy issue. Gore's outspoken position on climate change, symbolized by the publication of his book, *Earth in the Balance: Ecology and the Human Spirit,* was one of the reasons why Bill Clinton picked Gore as his vice-presidential running mate.

Clinton and Gore clearly seemed to be the dream team for climate change pol-icy. Clinton was basically willing to support binding limits on emissions and en-visaged passing energy taxes to link the reduction of greenhouse gases to reducing the budget deficit. In the end, he only managed a modest increase in the gasoline tax, which, together with the mishandled health reform proposals, contributed to the Republican takeover of Congress in the 1994 elections. From this point on-ward, it became clear to the Clinton administration that its political capital and room for maneuver in climate change policy was limited.

At the 1995 climate conference in Berlin, Clinton refused to make any com-mitments until he got past his reelection, meaning that any decisions were effectively postponed until the 1997 Kyoto climate summit. The Berlin meeting, however, made an important decision that was to heavily constrain American climate policy in the future: as an interpretation of the principle of "common, but differentiated responsibilities," it exempted developing countries, including the emerging market countries with strong economic growth such as China, India, and Brazil, from any obligations to cut emissions (Antholis et al. 30). This condition was to have major repercussions in the domestic debate on climate change. In July 1997, a few months before the Kyoto summit, a bipartisan group of senators sponsored a resolution that set conditions for the Senate ratification of any international climate change treaty: the minimization of any negative economic impacts and reversal of the Berlin decision to exempt developing countries from any obligations. The Byrd-Hagel Resolution, named after the pair of Republican and Democratic sponsors, passed by a margin of 95:0 (United States Senate). At the same time, industry groups stepped up their efforts to lobby against ghg limits as having high economic costs and undermining American competitiveness (Weart 2009a). The impact of Byrd-Hagel cannot be underestimated. More than any protestation of good intentions by administration policy-makers, Byrd-Hagel set the stage for the future course of American policy. It is no exaggeration to call it the basic law of American climate change policy, given that the Senate has to approve treaties, and both chambers have to pass any type of domestic climate change legislation.

The Clinton administration was severely limited in its climate change policy choices. When Clinton announced the American negotiating positions shortly before the Kyoto meeting, he pledged to return American emissions to 1990 levels by 2012—which still translated into a significant cut of 10 percent of the 1997 level, but fell well short of the original European target of a 15 percent cut to below 1990 levels. The final agreement was a cut of 7 percent for the United States and 8 percent for the European Union to below 1990 levels. By 1999, US emissions had grown by 12 percent and by 2008 by another 10 percent, which under the Kyoto commitments would have necessitated cuts of 30 percent by 2012, a target impossible to achieve (Victor 3-4). Clinton never submitted the Kyoto Protocol for ratification to the Senate (Antholis et al. 35-37). Attempts to implement the Kyoto Protocol with modifications acceptable to the United States failed, partly also because of European resistance to emission-trading (Victor 115). There was goodwill on the side of the Clinton administration, but no agreement was ratifiable domestically. It could well be argued that, given the high emissions growth during the boom years of the 1990s, American participation was doomed from the start. Only a massive purchase of emission rights from the former Soviet Union and other transition economies that had surplus emission rights due to the dismantling of Soviet industries would have made the United States Kyoto-compliant. But a transfer of roughly one hundred billion dollars to countries such as Russia would have been politically unacceptable and, in addition, would have amounted to the purchase of "hot air" (Victor 8-9).

Any attempts to maintain an accommodating stance on climate change policy by the incoming George W. Bush administration was quickly crushed by Vice President Cheney. The Bush administration finally buried the Kyoto Protocol, calling it "fatally flawed," and in rejecting it made explicit reference to the Byrd-Hagel Resolution (Knothe 311-20; Antholis et al. 38). A period of climate policy obstructionism or semi-isolationism set in, which only relented at the end of the Bush administration. The Bush administration's open rejection of Kyoto led to an outcry among Europeans and to major tensions in transatlantic relations. It laid open a transatlantic rift that clearly established the US as a laggard compared with other developed countries. The only bright spot in the US was increased activity on the subnational level, symbolized by the creation of an emissions trading system for power plants by ten Northeastern states and by the implementation of stricter standards for car emissions and a host of other measures in California (Rabe 2004, 2010a). In part, local and state activism was a reaction to the Bush administration's inactivity and obstructionism.

Despite dismay at the Bush administration's position, it could be argued that it was simply brutally honest about the ability of the American polity to ratify and deliver effective policy on the basis of the Kyoto Protocol, whereas the Clinton administration had proposed ambitious goals without ever detailing how they might gain legislative approval (Harrison 112). The Europeans turned out to have set themselves less ambitious goals than it appeared, as major European economies such as Germany had benefitted from the 1990 reference year; the demise of East German industry accounted for a large part of their reduction (Victor 5). Before the financial crisis hit, it also remained unclear whether many EU member states would reach their reduction targets (Posner et al. 62-65). The harsh stance of the Bush administration had some good sides, as it led to a reality check. It clearly helped to highlight the deficits of the Kyoto Protocol, such as the role of transition economies in emission trading and the failure to include emerging market economies, an issue that became more salient when China emerged as the largest emitter of CO_2 in 2008. But the fact remains that by 2005 the European Union had established a cap-and-trade system, whatever its shortcomings and limitations were, and was ready to move towards the next stage. In the United States, in contrast, legislative gridlock prevented any further movement towards meeting reduction targets. Which factors account for this state of affairs?

The Role of Public Opinion

One explanation for the laggard status of the United States is that climate change policy does not have enough support in the belief systems of Americans. The American public may not assign great priority to this issue and, due to historical experience and cultural framing, may simply be geared to a greater consumption and exploitation of natural resources while discounting and denying the possible consequences, as American historian David Potter pointed out more than fifty years ago. Public opinion polls on climate change do not confirm this thesis. Americans are aware of the perils of climate change and believe that it is a serious issue that requires government action (Woods Institute 2010b). However, there are deep political splits in the American public, a greater sensitivity to economic circumstance and costs, a greater receptivity to counter-arguments, and less engagement in the issue than in other OECD countries. Above all, even when there is support for government action in a general sense, climate change measures take a back seat on the agenda when competing with issues such as jobs, economic growth, or social policy.

Awareness of climate change among the American public has grown slowly. The watershed year was 1988, when polls after a summer heat wave and well-publicized congressional hearings showed a dramatic jump in awareness of the issue from less than 40 percent of the population in 1981 to almost 60 percent (Weart 2009b). By 2006, the awareness rate had risen to 91 percent. In polls conducted between 2006 and 2008, 70-77 percent of respondents acknowledged that the earth was warming. Between 1992 and 2007 there was also dramatic growth in the understanding of the phenomena: more than three-quarters of Americans claimed to have a reasonable grasp of the issue, up from a little more than half. By 2007, 41 percent ascribed climate change to anthropogenic causes, 42 percent saw a mixture of natural and anthropogenic causes. In 2008, 60 percent of respondents saw climate change as "a very serious issue," 32 percent as "somewhat serious" (Borick 27-31). Adding up these numbers would indicate the existence of an overwhelming consensus regarding the gravity of the issue. Critical in the development of this consensus were natural disasters, particularly Hurricane Katrina, high-profile events such as the screening of Al Gore's movie *An Inconvenient Truth*, and the IPPC's fourth assessment report, issued in 2007. All of these events found extensive coverage in the media (Weart 2009b).

However, there are still questions as to how much weight to assign to these polls, and how to interpret frequently vague attitudes. Digging deeper, fissures emerge that show conflicted attitudes about climate change. The deepest split is political, between Democrats and Republicans: while 80 percent of Democrats believe that there is solid evidence for climate change, less than half of Republican respondents share this belief. An almost identical difference exists with regard to the need for government action. The electorally crucial group of Independents are located between the two partisan groups, with over 60 percent acknowledging the problem of climate change. It is noteworthy that among this important swing group, the belief in the reality of climate change has shrunk by 13 percentage points since 2008 (Pew Research Center for the People & the Press 2009b).There are also regional differences, with much less support for climate change action in southern and southeastern states (Borick 32, 37; Rabe 2010a: 11). More importantly, polls taken in the past two years have registered increased skepticism about the seriousness and plausibility of climate change, a development that came to the fore even before the mail-hacking incident known as "Climategate" was widely publicized in December 2009. In 2009, only 51 percent believed that climate change was a "serious problem," down from 60 percent in 2008. The number of respondents who believed that their state had already felt negative effects of global warming fell from 28 percent to 16 percent. A poll also showed that between 2008 and 2009 there was a dramatic drop in the number of people who were likely to disagree with statements that discounted the evidence of climate change or implied that scientists and the media were overstating the risks associated with climate change (Borick 30, 34-36). The Climategate incident only added momentum to attitudes that were already shifting in a negative direction, and it is not quite clear whether all doubts in the public mind have been assuaged. Concerns about the uncertainty of climate change science and the tendency towards groupthink have not been removed and will continue to reverberate in public opinion in the US (Crook; Turnbull).

Equally controversial is the question of which specific measures the government should adopt. Beneath the veil of general consensus that something should be done, there is little or no consensus about what. While more than three quarters of respondents in a recent GFK-Roper poll favored government action to limit ghg emissions from businesses, almost the same proportion opposed higher gasoline taxes or taxes on electricity. In other polls, opposition to electricity taxes was found to be higher than 80 percent. Opposition to taxes is deeply rooted and evenly spread among the most important demographic and partisan groups (Woods Institute 2010a). While there is some support for the idea of paying for the production of renewable energy, most striking and ominous for the Obama administration is the waning support for cap-and-trade policies. Between 2008 and 2009, support for such policies dropped from 55 percent to 39 percent, while the proportion of those opposed reached 51 percent! The drop in support was even more dramatic when a price tag was associated with such policy and the price was shown to be increasing (Borick 53).

The bottom line is that there is general acknowledgement of the climate change issue and support for government action, but both awareness and support have been decreasing since 2008, which is probably due to the deep recession following the financial crisis. Majorities support modest regulatory action and increased use of renewable energy, but when it comes to policies that involve concentrated and direct costs such as taxes or higher charges, it turns out to be much more difficult to generate robust majorities. In addition, the highly visible partisan divide will hamper the necessary consensus building for far-reaching policy change. It is against this backdrop that the Obama administration's efforts to change the course of US climate change policies and the reactions to it have to be interpreted. In particular, the impact of the deep recession on voters' attitudes cannot be overestimated. It therefore comes as no surprise that when, on the eve of Obama's inauguration, the Pew Research Center asked Americans what their top political priorities were, climate change came in last in twentieth place.

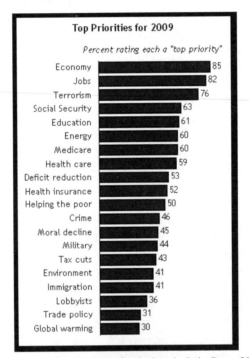

Source: Pew Research Center for the People & the Press, 2009a

Last but not least, any account of public opinion is incomplete without reference to the peculiarities of US discourse on climate change and of the way US media deal with the issue. A special factor in the United States is the existence of a small but active conservative think-tank scene that has consistently questioned the validity of mainstream climate change science, has stressed the costs in terms of reduced economic growth and job loss, and has depicted environmental activism as left-wing and directed against fundamental American values (Jacques et al. 354). The activities of these think tanks have laid a (somewhat questionable) academic foundation for climate change skepticism. With regard to the media, the American public is more receptive to climate change skepticism, because climate change skepticism finds greater resonance in the US media. That is true even of "balanced" mainstream media, but more importantly for the right-of-center media scene, such as the popular *Fox News* and radio talk show hosts such as Rush Limbaugh, and the elite *Wall Street Journal*, which all serve as a sounding board for climate-skeptical interpretations of events and polls (Media Matters; Starr). Public opinion plays a role in preventing the United States from adopting effective policies to curb ghg emissions and can thus help explain the US laggard status. The support for ghg reduction is not as strong and not as compelling as in most European countries, as international comparative polls show. In the 2010 Global Attitudes Survey, only Poland displays less concern about climate change than the United States among developed countries polled.

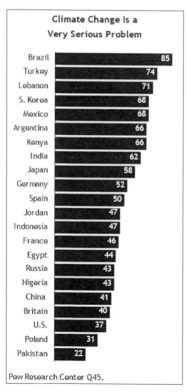

Source: Pew Research Center (70)

The Political System as an Obstacle to Climate Change Policies

The US political system and the resulting political processes also contribute to the difficulties of developing effective climate change policies. The winner-take-all electoral process has created a two-party system, which has practically barred third parties from arising except as a temporary voice of protest. For this reason, the green parties that play such an important role in many of Europe's proportional representation systems, and in influencing government policies on environmental issues, are not an effective force in the US party system. The institutionalization of climate change policies as an issue is difficult. Environmental concerns must fight their way into a party system that does not allocate a single party and its issues a firm space in the institutionalized political spectrum. A close association between issue and party, as with the green parties in Europe, is not possible. Also, the rules of representation make a difference. In the US Senate, they favor thinly populated, resource-rich states which are based on the exploitation of natural resources, agriculture, and automobile-based transportation systems, at the expense of densely populated, urbanized states on the east and west coasts that have been primarily, but not exclusively, the sources of support for environmental policies. The Senate is not representative and gives states with smaller, less environmentally friendly populations undue veto power. In addition, internal decisions rules in the Senate (cloture to end a filibuster) require sixty votes to produce policy change, thereby increasing the veto power of a small number of states. Theoretically, the twenty-one most sparsely populated states comprising merely 11.2 percent of the population can block any action (Sabato 146, 23-28). That Obama's climate change initiative floundered in the Senate is no surprise.

The separation of powers with independent electoral processes for the executive and the legislative branch obviate any need to support a government and put a premium on representing the electoral constituency; in the case of the Senate, the premium is on the individual state instead of the national interest or the nationwide conception of an issue such as climate change (Sabato 19, 22). Instructive is the Canadian case, where party discipline allowed the ratification of the Kyoto Protocol despite opposition in the Cabinet and from provincial governments (Harrison 112). The overemphasis on particularized geographic interests fragments the policy process and makes access for powerful, well-funded business lobbies easier, particularly when geographic and functional interests overlap, as for example in the case of the coal state West Virginia. This feature of the system also accounts for the lack of focus in climate change bills that distract from the basic purpose of emission reductions with a plethora of special provisions, exemptions, and subsidies that satisfy special and geographic interests and dilute the effectiveness and transparency of legislation (Rabe 2010a: 7, 18). Congress seems inept in addressing long-term programs with intergenerational consequences such as entitlement reform or energy, and climate change clearly falls into this category.

In fact, environmental legislation has stalled since the early 1990s, ever since the passage of the amendments to the Clean Air Act. Only eight major initiatives were passed between the 103rd and the 110th Congress, while thirty-four were enacted between the 91st and the 102nd. Congress has failed to overhaul and modernize the legal basis for environmental policy (Klyza et al. 94). One of the reasons for this situation is increasing ideological and partisan polarization, which has also trapped the climate change issue and made it politically intractable (Nivola et al.). Mainstream Republicans are in fundamental opposition to climate change policy, but also sectoral and regional fault lines rebel against it. As a consequence, energy and climate change legislation has been of the lowest common denominator type, with an emphasis on subsidies, veiled regulation, or extreme regulatory discretion (Rabe 2010b: 262-65). A complement to congressional avoidance of substance on this issue is the growth of political grandstanding and attention grabbing through the proliferation of hearings. Two hundred hearings were held in 2007-2008 alone, while only 175 hearings were held between 1975 and 2006. Congress feigned attention, but failed to develop serious deliberation of the issue to arrive at coherent policy development. Balkanized committee jurisdiction, particularly in the Senate, encouraged such an approach. A number of climate change bills were introduced, most prominently the McCain-Lieberman Bill, but in the Senate, except for procedural cloture votes, no bill ever came up for a vote on the floor (Rabe 2010b: 267-72; Brewer 2010).

In the end even ACES, the House's successful attempt at climate change legislation, suffered from the shortcomings of the legislative process in that it turned serious climate change policy into distributive policy by greatly curtailing the use of auctioning of permits, something which President Obama had originally advocated, and using instead free allocations as distributional tools, on top of subsidies, "offsets" for the agricultural sector, and various mandates and standards favoring particular constituencies to buy support for the bill. The basic goal was to turn the legislation into distributional policy that would camouflage its true impact on energy prices, or postpone any such impact to a later date. This made reliable cost analysis virtually impossible, and in the end made the bill vulnerable to claims by opponents that it would impose massive costs on consumers and industry, which influenced the chances of passage in the Senate (Rabe 2010b: 278-82; Congressional Research Service 2009b). In addition, it was not at all clear whether the United States possessed the administrative capacity to implement comprehensive climate change policy, when grafted onto a fragmented, dysfunctional management structure. Complex climate change policies require massive horizontal and vertical coordination and additional resources and structures for those agencies charged with carrying out the mandate. The capacity of the EPA as the presumed lead agency was disputed by some, and the division of responsibilities among other agencies, such as the Departments of Energy, Commerce, and Agriculture, was problematic and not fully resolved. Congress left some fundamental governance issues such as funding, staffing, and technical and economic expertise unaddressed (Rosenbaum 296-302).

The Influence of Business and Competitiveness Issues

That the American political system, particularly the US Congress, gives multiple access points for organized interests has already been noted. That is particularly true for business interests, especially if they are concentrated on a regional basis. And one way to construe the laggardness of the US is to attribute it to undue business influence or influence that is not sufficiently counterbalanced by environmental groups or political actors, such as political parties. What is striking in the US as compared with European economies is that there was organized business opposition against climate change mitigation policies and the Kyoto Protocol in particular, supported by an outright denial of the mainstream scientific position on climate change. Such sentiments were powerfully voiced by the Global Climate Coalition (GCC), an association of large energy, automobile, and other manufacturing firms opposed to effective ghg reduction. Its undisputed leader was Exxon-Mobil. The GCC is credited with convincing the George W. Bush administration not to ratify the Kyoto Protocol. The American Petroleum Institute has also been consistently opposed to mitigation measures and in 2001 even questioned a report by the US National Academy of Sciences confirming recent evidence for global warming (Brewer 2009: 62, 65). A significant force militating against ghg reductions, again setting the US apart from European countries, is the small business sector, which is effectively represented in the US Chamber of Commerce and in ad-hoc groups such as the Small Business Survival Committee. Their influence is enhanced by the fact that they have a nationwide network that has influence in every congressional district.

However, the anti-ghg reduction business phalanx is not monolithic, nor did the GCC turn out to have staying power. A number of large corporations such as Ford, Shell, BP, Texaco, and Daimler have subsequently left. The subset of subsidiaries of European corporations that operate in different political environments in their home base, where they support ghg reduction, came under even stronger pressure to leave. Responses of the business sector to regulation of emissions also vary according to the degree to which their revenues or expenses are dependent on fossil fuel prices. This is true for the oil, coal, auto, steel, and aluminum industries. Others, such as the insurance industry, which may suffer from losses due to climate change-induced weather irregularities, support action on climate change, as do companies that would benefit from fostering renewable energy and a more efficient energy infrastructure (Brewer 2009: 62-63). During the G.W. Bush administration, there was also a shift in the business sector, when companies that were previously opposed to any type of climate change policy altered their course because they anticipated government intervention and preferred regulatory certainty for their future investments.

That the GCC eventually fell apart is also a reflection of the fact that sustained opposition to climate change by large corporations was not good publicity in view of increasing public concern and state government activism. This concern led companies such as GE, Alcoa, and Duke Energy to join the US Climate Action Partnership (USCAP), an advocacy coalition of five leading environmental groups[2] and twenty-six major corporations (among them Ford, Shell, Siemens, NRG Energy, and Rio Tinto) that support a moderate ghg reduction regime based on a cap-and-trade model. USCAP is credited with formulating the blueprint for ACES and inducing the necessary deals and bargains to allow passage of the bill in the House (Rabe 2010b: 279-81; US Climate Action Partnership). However, the fate of the climate change bill in the Senate underlines the limits of moderate climate change activism by large multinational corporations. A formal business organization dedicated to thwarting any climate action was not necessary. Opposition could conveniently be left to coal interests and those members of Congress representing coal-producing or coal-using states, including ten Democratic senators from states such as Ohio, which relies on coal for 86 percent of its electricity generation (Broder). Conservative sentiment, opposition from coal states and small businesses, and fear of higher energy and gas prices for consumers weighed more strongly than the influence of forward-looking corporate engagement from organizations such as US-CAP.

Another factor aligned with business opposition to climate change action was that of competitiveness concerns. Although a number of studies have shown that a modest ghg tax (or equivalent measure) would have minimal impact on US industry, loss of competitiveness remains a major concern (Aldy et al.). It is no coincidence that domestic resistance to effective policy change coincided with the rise of emerging market economies, such as China, India, and Brazil, which are seen as international competitors, but are not subject to ghg control regimes, and thus in the estimation of business would gain additional competitive advantages in not having to fulfill costly ghg reduction requirements. The focus here is primarily on

---

[2]   The five groups are the Natural Resources Defense Council, the Environmental Defense Fund, The Nature Conservancy, The World Resources Institute, and the Pew Research Center on Climate Change.

China, with which the United States has a sizeable trade and current account deficit (Bergsten; Cline) and which is held responsible by Congress and unions for major US job losses in manufacturing, gaining advantages because of the absence of climate change policy and lax enforcement of environmental regulations. The permanent concern about US competitiveness—imagined or real—puts the brakes on climate change activism and serves as an argument by business lobbies, unions, and those associated with them to discourage any measures that would exacerbate the US competitive position vis-à-vis emerging market countries. The high saliency of competitiveness concerns that are fueled by news about the trade deficit and loss of jobs to emerging market countries frames the debate about climate change (Houser). The public in general is much more receptive to such arguments than in countries such as Germany, which has had a substantial trade surplus over a long period.

This is also one of the reasons that ideas on how to level the playing field through trade-restrictive measures were first aired in the United States, and it also explains why every climate change bill in the United States so far has contained border measures that would put a tariff on energy-intensive goods from countries that do not undertake climate change mitigation efforts or requires importers to buy emission allowances, which raises the issue of WTO compatibility (Brewer 2003; Hufbauer et al. 2009). This implies that, from a competitiveness perspective, the US maintains high reciprocity expectations when entering international climate change agreements. The salience of this issue explains why every piece of climate change legislation will contain trade measures to assuage American competitiveness concerns, and why the US is likely to push for the legalization of such measures on an international level. Opposition to climate change measures by business can take many forms and avenues, but it is crucial in the US context that some arguments, such as competitiveness, can be linked to more general concerns that resonate strongly with the larger public and with non-business groups.

Disaffection with UN Climate Change Diplomacy

US participation in climate change agreements is hobbled by unease over entering into international commitments and ceding sovereignty to international organizations. For many issues requiring international cooperation, there is a strong urge in the US to go it alone or at least not enter new agreements (Luck), demonstrating ambivalence about international agreements and multilateral institutions. This ambivalence is most evident in the fact that the United States has ditched the Kyoto Protocol, but has continued to work within the UNFCCC to seek a successor to it. It is also evident in a US strategy pursued by the Bush and the Obama administrations to seek alternative, but parallel negotiating venues outside the UN process, such as the Major Economies Forum, which brings together the sixteen leading CO_2-emitters, or as a somewhat larger, but similar alternative, the G-20. The reason for pursuing this alternative track is that the UN process, with its 193 members and its policy of consensus is seen by some as unwieldy and inefficient. The highly visible UN proceedings have also been perceived as encouraging false hopes and exposing the US to attacks by singling the nation out as the sole force responsible for blocking progress (Antholis et al. 69).

In the American view, the UN process also tackles too many issues and aspects of climate change at a single time and in a single forum. It gives undue influence to developing countries, who see the UN process as a way to extract additional aid and convert the process into a massive redistribution of resources to developing countries, while failing to allow for a crucial differentiation between developing countries and high-growth emerging market countries, which would give a more realistic rendition of the principle of common and differentiated responsibilities. It does not really make sense to treat China and Chad alike; as a matter of fact it leads to unholy alliances (Green et al. 3-4). It also rewards "spoilers" like oil-exporting countries (Saudi Arabia) who want to be compensated for losses of oil revenues, and "snipers" such as Venezuela, Bolivia, Sudan, Iran, and Cuba, who undermine consensus for purposes of political grandstanding and principled opposition to the United States. All of these countries, from the US perspective, distract from rather than contribute to the goal of emission reduction (Antholis et al. 51-54; Hufbauer et al. 2009; Posner et al. 73-75, 123-26). The US government is also increasingly disaffected by the disorderly decision-making and chaotic procedures at UN climate summits, as Secretary of State Clinton hinted when she described the Copenhagen summit as "the worst meeting I've been to since eighth-grade student council" (Antholis et al. 64).

US distrust in the UN process is fuelled by the lack of enforcement mechanisms and the fear that negotiating partners will simply shirk their responsibilities and not comply with their treaty commitments, while ratification in the United States would enshrine commitments in national law (Goldsmith et al. 213, 216-17). The UNFCCC simply lacks the administrative capacity and authority to enforce compliance with targets and timetables, and does not even provide for effective rules for carbon accounting (Green et al.). In the context of climate change policy, compliance, originally assumed to be merely a technical issue, has been transformed into a strategic issue, with emphasis shifting to monitoring, reporting, and verification (also called MRV) of emissions. Not only is the value of any international agreement undermined if emissions are not effectively monitored, correctly reported, and independently verified, but the smooth operation of an international emission trading system would depend on these mechanisms. The refusal of emerging market and developing countries to submit even their voluntary commitments to independent verification bode ill for greater American engagement

(Hufbauer et al. 2010). Currently, emerging market countries have only limited obligations to report greenhouse gas inventories and information on mitigation policies, resulting in patchy data from some of the world's largest emitters. Accuracy, transparency, and comprehensiveness of emissions monitoring and accounting is totally inadequate and undermines the legitimacy of the UNFCCC process (Green et al. 9-10). Until there is improvement in these areas, the United States will be reluctant to make firm and far-reaching commitments.

In the absence of a compliance mechanism, the collective target and timetable approach pursued under Kyoto does not seem to be working. In actual negotiations, countries offer targets from the bottom-up that are based on their calculation of domestic and international political interests. Coupled with effective compliance and measuring mechanisms, a new bottom-up framework for climate policy to be established outside, but parallel to the UN-process would probably be more acceptable to the US and possibly also to many other major emitters from developing or emerging market countries. In spite of its serious limitations, there are elements in the Copenhagen accord that point in this direction, though its full institutional implementation is still some time off (Green et al. 7-10). This does not mean that the United States will ditch the UNFCCC process, or abandon support of the poorest countries, but that the real reduction negotiations will happen by means of a parallel track, including all major emitters, and taking a different approach to the one agreed on in the Kyoto Protocol. The insistence on effective compliance mechanisms is certainly useful, but the United States will hardly lose its laggard status if it uses the incompleteness of instruments and processes as an excuse for inaction or the lack of credible reduction commitments of its own.

Summary: Is There a Way Forward?

All four of the factors identified here conspire to relegate the United States to a laggard status, particularly compared to Europe. It is hard to see whether there can be major changes in any of these dimensions in the near future. In public opinion polls, support for the validity of climate change scenarios and appropriate action are in decline, even though there is still majority support for some action. The protestations of support may actually only signal a commitment to action that is not very deep or intense, particularly when voters are confronted with actual costs such as higher taxes or the costs of non-compliance by other states, should the latter be clearly identified and brought to the attention of voters (Goldsmith et al. 216-17). The current political and economic environment does not bode well for a change in public opinion. The deep recession has put climate change issues on the back burner, has made issues of job creation and growth paramount, and has created a backlash against policies that can be construed as leading to higher energy costs.

As studies have shown that recessions after financial crises usually last longer, and lead to lower GDP growth and higher unemployment in the decade following the crisis (Reinhardt et al.), the economic context for climate change policy after the deep financial crisis does not look favorable.

On the political side, the backlash is in full swing, with Republicans denouncing a cap-and-trade system as just another tax during the 2010 campaign. What the 111th Congress achieved on climate change, with a diluted cap-and-trade-bill passed in the House, but no action in the Senate, is the maximum of what could have been achieved given the strictures of the US political system. And since Republicans, driven by the virulently anti-tax, anti-government, grassroots Tea Party movement, managed to recapture the House in the 2010 mid-term election, winning sixty-three seats in the House, and six in the Senate (Balz et al.; Luce), Congressional action on climate change is barred at least for two more years, if not longer. The Republican majority in the House after the 2010 mid-term elections is likely to lead to increasing ideological polarization, and, as climate change serves as a wedge issue between Democrats and Republicans, mobilization against major climate change initiatives is likely. Within the Republican caucus, climate change skepticism has moved into the mainstream again. The majority in the Republican caucuses in both chambers belong to a group that the Center for American Progress has dubbed "climate zombies": 76 percent of the Republicans in the Senate and 52 percent of Republican house members have publicly questioned mainstream climate change science. Eighty of the eighty-seven Republican freshmen in the House have signed pledges that they would resist any action on climate change (Johnson). Mobilization against global warming policies was one of the items on the Tea Party agenda, albeit not the most prominent one, as the focus was on the big government theme. According to polls, 70 percent of Tea Party adherents see no solid evidence that climate change is occurring, 75 percent believe it is not a serious problem, and only 8 percent support any government action (Pew Center for the People & the Press 2010). The few remaining moderate Republicans, such as Susan Collins and Olympia Snowe (both from Maine), and Senator McCain (Arizona), who until Obama's victory supported ghg reductions (prior to his own candidacy for the presidency was the first Senate Republican to hold hearings and sponsor a bill on this issue (Drew 127-28)), will most likely be completely marginalized or will abandon their support for climate change mitigation measures, as McCain has done in the fight to save his Arizona Senate seat.

It is highly likely that, given the Republican surge, the EPA action to classify CO_2 a pollutant to be regulated will produce a major political conflict between the Obama administration and the Republicans over the latter's attempt to rescind the EPA's authority (Congressional Research Service 2009a; New York Times 2010c). Rescission is unlikely to happen, and would be blocked in the Senate or even by a presidential veto, but even moderate Democrats in the Senate are trying to rein in the EPA by proposing that it postpones all regulatory action for another two years.

Climate change policy will basically resemble that of the Clinton era, with the visible obstructionists now located in Congress and the administration trying to develop a national ghg reduction strategy under severe constraints. Democrats (and the Obama administration) may also try once again to frame climate change measures as job bills or as energy independence bills. And Democrats are most likely to respond to Republican obstructionism by sponsoring hearings highlighting the risks of inaction. The emphasis will be on putting Republicans on the defensive, but without engaging in a substantive debate, for instance over the benefits of cap-and-trade vs. a carbon tax. In the context of a deep recession, a name-and-shame strategy by Democrats may not have much resonance with voters and the general public. There may be one other outcome of the approaching Congressional debate about climate change policy: if Republicans are successful in denouncing cap-and-trade as a veiled tax (New York Times 2010a) and thus transfer the aversion of carbon taxes to a cap-and-trade approach, thereby negatively leveling the playing field, it may eventually be possible for advocates of a tax to shift the debate to price-type approaches to climate change abatement (of which a carbon tax is a variant), which many economists see as an instrument both more effective and easier to administer (Nordhaus 2008a: 148-64).

The role of the business community may be less visible than previously, when the most conservative groups tried to undermine the validity of climate change science or the Kyoto Protocol. Businesses opposed to ghg reductions will not have to engage in fundamental fights. With Republicans having a solid majority in the House, and an effective veto in the Senate, there is little to worry about regarding new climate change legislation. Businesses will most likely focus on EPA regulation of greenhouse gases. Here, groups such as the US Chamber of Commerce, the National Association of Manufacturers, and coal producers and utilities using coal will likely come out in support of curtailing the EPA's mandate. The interesting question is how business groups such the US Climate Action Partnership, which supported a cap-and-trade-system, will position themselves.

As far as the UN-based diplomacy process is concerned, it is most likely that the United States will continue to work within the UNFCCC, but put equal emphasis on alternative forums such as the Major Economies Forum. Its approach will be to build on the evolution of the bottom-up approach, consistent with the Copenhagen accord. The administration will rhetorically maintain the 17 percent reduction by 2020 that is enshrined in ACES, to be reached by EPA regulation and other action. It will insist on the inclusion of and effective action by emerging market countries, particularly China, India, and Brazil, not the least in order to shift sentiment in an increasingly skeptical Congress. There will be renewed insistence on construction of solid and robust systems of monitoring, reporting, and verification

(MRV). Eventually, there may also be more support for a carbon tax which poses fewer problems in terms of honesty, transparency, and effective administration (Nordhaus 2008b). In any event, the United States will drift away from the architecture of the Kyoto Protocol and most likely work towards a post-Kyoto system, whose outlines must remain vague.

If there is stalemate at the federal level, it is most likely that greater action will shift to the state and local level, and that state emission trading schemes such as the Regional Green House Gas Initiative (in ten northeastern states) and the Western Climate Initiative (in western US and Canadian provinces) will come into focus (Raymond 108-13). But decentralized approaches can be claimed to lead to inefficient, patchwork solutions that are uncoordinated, cost-ineffective, and mutually undermining, a proposition that is not appealing to businesses operating beyond jurisdictional borders (Antholis et al. 40; Posner et al. 68-69). At the same time, local and state initiatives may be as much circumscribed by economic conditions as on the federal level. In California, however, the lead state for climate change and environmental initiatives, a referendum which sought to postpone Arnold Schwarzenegger's climate change initiative until unemployment drops substantially in the state, was defeated by a margin of 61 percent to 39 percent. This seems to indicate that on the regional level there is still scope for action, even under adverse economic conditions (Brewer 2010; California Secretary of State).

There remains one incalculable element: events. A major environmental disaster on the scale of Hurricane Katrina, or the experience of fundamentally adverse weather conditions such as a major drought, whether they are attributable to climate change or not, may sway public opinion and influence the policy debate. Such events shape perception of and attitudes towards the phenomenon of climate change (Borick 31-36). The oil spill in the Gulf of Mexico in April 2010 led to a significant shift in public opinion over the priority to be accorded to environmental protection over energy production. In a Gallup poll taken in May 2010, 55 percent gave priority to the protection of the environment, even at the risk of limiting the amount of energy supplies, up from 43 percent in March. The change occurred only among Democrats and Independents, though. The spill reversed a longer-term trend of favoring economic growth over environmental protection (Gallup). Reports suggest that the immediate and medium-term impacts of the Macondo well spill are less severe than the Exxon-Valdez disaster (Kekulé).

The fact remains that the United States has not been able to establish a national ghg reduction regime under an activist Democratic president and Democratic control of both houses of Congress, and it is highly uncertain whether it will even achieve this aim in the medium term. The laggard status thus appears confirmed. It can be argued, however, that this is a very Eurocentric perspective. The European Union can champion a (relatively well) functioning trading system, which has given it valuable experience in designing and operating an emissions trading system. But currently, the EU system covers only 44 percent of total EU CO_2 emissions and only 8 percent of global emissions (Congressional Research Service 2010; Nordhaus 2008b: 92). In the post-Kyoto era, the US position is not that far from that of countries such as Canada, Australia, Korea, and possibly Japan. The US emphasis on a bottom-up approach for a post-Kyoto system, and the intensive attempts to engage emerging market countries as well as a push for the construction of a solid MRV system may define a core agenda for a US-led coalition of developed and emerging market countries pursuing a different course of action than that of the European Union. In Copenhagen it was the European Union that was isolated when the major deals were struck (Antholis et al. 63-68).

So far, the United States may be lagging just behind Europe. The real downside of the domestic paralysis in the United States and the lesson of the first two years under Obama is that the United States is losing valuable time in constructing a domestic ghg system which critically needs trial and error to get it running. The laggard marker is also a useful reminder that without a credible US commitment to reduce greenhouse gases, which eventually has to be reflected in national legislation, other major emitters such as China and India, which the United States sees as essential parts of a global climate change regime, will not move. The ball remains as much as ever in the US court.

Works Cited

Agrawala, Shardul. "Context and Early Origins of the Intergovernmental Panel for Climate Change." *Climatic Change* 39.4 (1998): 605-20.

Aldy, Joseph E. and William A. Pizer. *The Competitiveness Impacts of Climate Change Mitigation Policies*. Arlington, VA: The Pew Center on Climate Change, 2009.

Antholis, William and Strobe Talbott. *Fast Forward. Ethics and Politics in the Age of Global Warming*. Washington, DC: Brookings Institution, 2010.

Balz, Dan and Jon Cohen. "Republicans Making Gains Against Democrats Ahead of Midterm Elections." *The Washington Post* 7 September 2010. Accessed 7 September 2010. <http://www.washingtonpost.com/wp-dyn/content/article/2010/09/07/AR2010090700007.html>.

Bergsten, C. Fred (ed.). *The Long-Term International Economic Position of the United States*. Washington, DC: Peterson Institute for International Economics, 2009.

Borick, Christopher P. "American Public Opinion and Climate Change." *Greenhouse Governance*. Ed. Barry G. Rabe. Washington, DC: Brookings Institution, 2010. 24-57.

Brewer, Thomas L. *The Political Economy of US Government, Business and Public Responses to Climate Change*, forthcoming.

———. "U.S. Government Policymaking on Climate Change: Recent Developments, Transitions, and Prospects for the Future." *Oxford Energy and Environment Comment, October 2010*. <http://www.usclimatechange.com/downloads.php/oies-cs_comment_tom_brewer_us_policymaking%255B1%255D.pdf>.

———. "Pluralistic Politics and Public Choice: Theories of Business and Government Responses to Climate Change." *Theory and Practice of Environmental Foreign Policy*. Ed. Paul G. Harris. London, New York: Routledge, 2009. 57-73.

———. "The Trade Regime and the Climate Regime: Institutional Evolution and Adaptation." *Climate Policy* 3 (2003): 329-41.

Broder, David M. "Geography Is Dividing Democrats Over Energy." *New York Times* 26 January 2009. Accessed 16 September 2010. <http://www.nytimes.com/2009/01/27/science/earth/27coal.html>.

California Secretary of State, Semi-Official Election Results, Tuesday, November 2, 2010, Proposition 23. Suspend Air Pollution Control Law (AB 32). Accessed 10 November 2010. <http://vote.sos.ca.gov/maps/ballot-measures/23/>.

Cline, William R. "Renminbi Undervaluaton, China's Surplus, and the US Trade Deficit." 2 August 2010. *Peterson Institute for International Economics*. Accessed 16 September 2010. <http://www.iie.com/publications/pb/pb10-20.pdf>.

Congressional Research Service. *Climate Change and the EU Emissions Trading Scheme (ETS): Looking to 2020. Report for Congress*. Washington, DC: Congressional Research Service, 2010.

———. *Climate Change: Potential Regulation of Stationary Greenhouse Gas Sources Under the Clean Air Act. Report for Congress*. Washington, DC: Congressional Research Service, 2009a.

———. *Greenhouse Gas Legislation: Summary and Analysis of H.R. 2454 as Reported by the House Committee on Energy and Commerce. Report for Congress*. Washington, DC: Congressional Research Service, 2009b.

Crook, Clive. "Climategate and the Big Green Lie." *The Atlantic Monthly* 14 July 2010. Accessed 27 September 2010. <http://www.theatlantic.com/politics/archive/2010/07/climategate-and-the-big-green-lie/59709/>.

Drew, Elizabeth. *Citizen McCain*. New York: Simon & Schuster, 2008.

Gallup. "Oil Spill Alters Views on Environmental Protection." 27 May 2010. *Gallup Poll*. Accessed 18 September 2010. <http://www.gallup.com/poll/137882/oil-spill-alters-views-environmental-protection.aspx>.

Goldsmith, Jack L. and Eric A. Posner. *The Limits of International Law*. New York: Oxford University Press, 2005.

Green, Fergus, Warwick McKibbin, and Greg Picker. "Confronting the Crisis of International Climate Policy." July 2010. *Lowy Institute.* Accessed 23 July 2010. <http://www.lowyinstitute.org/Publication.asp?pid=1329>.

Harrison, Kathryn. "The Road not Taken: Climate Change Policy in Canada and the United States." *Global Environmental Politics* 7.4 (2007): 92-117.

Houser, Trevor. "Green and Mean: Can the New US Economy be both Climate-Friendly and Competitive? Testimony before the Commission on Security and Cooperation in Europe, US Congress." 10 March 2009. *Peterson Institute for International Economics.* Accessed 1 September 2010. <http://www.petersoninstitute.org/publications/papers/houser0309.pdf >.

Hufbauer, Gary C. and Jisun Kim. "After the Flop in Copenhagen." March 2010. *Peterson Institute for International Economics.* Accessed 30 August 2010. <http://www.petersoninstitute.org/publications/pb/pb10-04.pdf>.

————. "The World Trade Organization and Climate Change: Challenges and Options." September 2009. *Peterson Institute for International Economics.* Accessed 17 September 2010. <http://www.petersoninstitute.org/publications/wp/wp09-9.pdf>.

Jacques, Peter J., Mark Freeman, and Riley E. Dunlap. "The Organisation of Denial: Conservative Think Tanks and Environmental Skepticism." *Environmental Politics* 17.3 (2008): 349-85.

Johnson, Brad. "The Climate Zombie Caucus Of The 112th Congress." *The Wonk Room.* Accessed 23 November 2010. <http://wonkroom.thinkprogress.org/climate-zombie-caucus/>.

Kekulé, Alexander S. "Ein Ende, das zu schnell aufatmen lässt." *Der Tagesspiegel* 23 September 2010: 8.

Klyza, Christopher McGrory and David Sousa. *American Environmental Policy 1990-2006: Beyond Gridlock.* Cambridge, MA: MIT Press, 2008.

Knothe, Danko. *Macht und Möglichkeit.* Berlin: LIT Verlag, 2007.

Luce, Edward D. "Tea Party Draws on Tradition of Activism." *Financial Times* 17 September 2010: 3.

Luck, Edward C. "American Exceptionalism and International Organization: Lessons from the 1990s." *US Hegemony and International Organizations.* Ed. Rosemary Foot, S. Neil McFarlane, and Michel Mastanduno. New York: Oxford University Press, 2003. 25-49.

Media Matters for America. "Fox News Fiddles with climate change polling." 8 December 2009. *Media Matters for America.* Accessed19 September 2010. <http://mediamatters.org/blog/200912080002>.

New York Times. "Climate and Energy Legislation." *New York Times* 23 July 2010a. Accessed 19 September 2010. <http://topics.nytimes.com/top/news/business/energy-environment/climate-and-energy-legislation/index.html>.

————. "Democrats Call Off Climate Bill Effort." *New York Times* 22 July 2010b. Accessed 14 September 2010. <http://www.nytimes.com/2010/07/23/us/politics/23cong.html>.

————. "Senators Want to Bar E.P.A. Greenhouse Gas Limits. *New York Times* 21 January 2010c. Accessed 27 September 2010. <http://www.nytimes.com/2010/01/22/science/earth/22climate.html>.

Nivola, Pietra S. and William A. Galston. "Delineating the Problem." *Red and Blue Nation? Consequences and Corrections of America's Polarized Politics. Vol. 1.* Ed. Pietro S. Nivola and David W. Brady. Washington, DC: Brookings Institution, 2008. 1-47.

Nordhaus, William. *A Question of Balance. Weighing the Options on Global Warming Policies.* New Haven: Yale University Press, 2008a.

————. "Is There Life After Kyoto?" *Global Warming. Looking Beyond Kyoto.* Ed. Ernesto Zedillo. Washington, DC: Brookings Institution, 2008b. 91-100.

Pew Center on Climate Change. "At a Glance. American Clean Energy and Security Act of 2009." 26 June 2009. *Pew Center on Climate Change.* Accessed 13 September 2010 <http://www.pewclimate.org/docUploads/Waxman-Markey-short-summary-revised-June26.pdf>.

Pew Research Center. "22-Nation Pew Global Attitudes Survey." 17 June 2010. *Pew Research Center Global Attitudes Project.* Accessed 25 September 2010. http://pewglobal.org/files/pdf/Pew-Global-Attitudes-Spring-2010-Report.pdf.

Pew Research Center for the People & the Press. "Wide Partisan Divide Over Global Warming. Few Tea Party Republicans See Evidence." 27. October 2010. *Pew Research Center for the People & the Press.* Accessed 27 November 2010. <http://pewresearch.org/pubs/1780/poll-global-warming-scientists-energy-policies-offshore-drilling-tea-party>.

————. "Economy, Jobs Trump All Other Policy Priorities In 2009." 22 January 2009a. *Pew Research Center for the People & The Press.* Accessed 15 September 2010. <http://pewglobal.org/files/pdf/Pew-Global-Attitudes-Spring-2010-Report.pdf>.

————. "Fewer Americans See Solid Evidence of Global Warming." 22 October 2009b. *Pew Research Center for the People & the Press.* Accessed 15 September 2010. <http://people-press.org/report/556/global-warming>.

Pommerance, Rafe. "The Dangers from Climate Warming: A Public Awakening." *The Challenge of Global Warming.* Ed. Dean Edwin Abrahamson. Washington, DC: Island Press, 1989. 259-69.

Posner, Eric A. and David Weisbach. *Climate Justice.* Princeton: Princeton University Press, 2010.

Potter, David M. *People of Plenty. Economic Abundance and the American Character.* Chicago: Chicago University Press, 1958.

Rabe, Barry G. "The Challenges of U.S. Climate Governance." *Greenhouse Governance. Addressing Climate Change in America*. Ed. Barry G. Rabe. Washington, DC: Brookings Institution, 2010a. 3-23.

———. "Can Congress Govern the Climate?" *Greenhouse Governance*. Ed. Barry G. Rabe. Washington, DC: Brookings, 2010b. 260-86.

———. *Statehouse and Greenhouse*. Washington DC: Brookings Institution, 2004.

Raymond, Leigh. "The Emerging Revolution in Emission Trading Policy." *Greenhouse Governance*. Ed. Barry G. Rabe. Washington, DC: Brookings Institution, 2010. 101-25.

Reinhardt, Carmen M. and Vincent R. Reinhardt. "After the Fall." September 2010. *National Bureau of Economic Research*. Accessed 17 September 2010. <http://www.nber.org/papers/w16334>.

Rosenbaum, Walter. "Greenhouse Regulation: How Capable is EPA?" *Greenhouse Governance*. Ed. Barry G. Rabe. Washington, DC: Brookings Institution, 2010. 286-310.

Sabato, Larry J. *A More Perfect Union*. New York: Walker & Co., 2007.

Starr, Paul. "Governing in the Age of Fox News." *The Atlantic Monthly* January/February 2010. Accessed 10 September 2010. <http://www.theatlantic.com/magazine/archive/2010/01/governing-in-the-age-of-fox-news/7845/>.

Turnbull, Andrew. "A Climate Overhaul Is Needed To Win Back Trust." *Financial Times* 26 September 2010: 11.

United States. Senate. "Byrd-Hagel Resolution." 25 July 1997. 105th Congress, S. Res. 98. Accessed 13 September 2010. <http://www.nationalcenter.org/KyotoSenate.html>.

US Climate Action Partnership. List of members. *U.S. Climate Action Partnership*. Accessed 15 September 2010. <http://www.us-cap.org/>.

US Supreme Court. "Massachusetts vs. EPA." 2 April 2007. 549 U.S. 497 (2007). Accessed 25 September 2010. <http://www.supremecourt.gov/opinions/06pdf/05-1120.pdf>.

Victor, David G. *The Collapse of the Kyoto Protocol and the Struggle to Slow Global Warming*. Princeton: Princeton University Press, 2001.

Washington Post. Live Results. Senate, House, Governors Races, Elections Maps, November 2010. <http://www.washingtonpost.com/wp-srv/special/politics/election-results-2010/>.

———. Votes Data Base, House Vote 477, June 2009. Washington Post. Senate, United States. "Byrd-Hagel Resolution." 25 July 1997. 105th Congress, S. Res. 98. Accessed 13 September 2010. <http://projects.washingtonpost.com/congress/111/house/1/votes/477>.

Weart, Spencer. "Government: The View from Washington, DC." June 2009a. *The Discovery of Global Warming*. American Institute of Physics. Accessed 13 September 2010 <http://www.aip.org/history/climate/Govt.htm>.

————. "The Public and Climate Change." July 2009b. *The Discovery of Global Warming.* American Institute of Physics. Accessed 13 September 2010. <http://www.aip.org/history/climate/public2.htm#M_105_>.

White House. "Remarks of President Barack Obama—As Prepared for Delivery." 24 February 2009. Address to Joint Session of Congress, 24 February 2009. Accessed 27 September 2010. <http://www.whitehouse.gov/the_press_office/ Remarks-of-President-Barack-Obama-Address-to-Joint-Session-of-Congress/>.

Woods Institute for the Environment. "Global Warming Poll." 9 June 2010a. *Woods Institute for the Environment.* Accessed 15 September 2010. <http://woods.stanford.edu/docs/surveys/Global-Warming-Survey-Selected-Results-June2010.pdf>.

————. "Large Majority of Americans Support Government Solutions to Address Global Warming." 9 June 2010b. *Woods Institute for the Environment.* Accessed 15 September 2010. <http://woods.stanford.edu/research/americans-support-govt-solutions-global-warming.html>.

Risk, Space, and Natural Disasters: On the Role of Nature and Space in Risk Research

Heike Egner

Introduction: Natural Catastrophes Are Not Natural

At a first glance, 'natural' disasters, that is to say, catastrophes induced by natural processes such as avalanches, floods, landslides, earthquakes or volcanic eruptions, seem to be 'natural,' inevitable, and therefore inescapable. On the other hand, catastrophes induced by societal processes or by human actions appear to be avoidable, controllable, or at least manageable. One might think of catastrophes resulting from the use of technology, such as the major incident with the offshore oil platform *Deepwater Horizon* in the Gulf of Mexico in 2010, or of a terrorist attack, or the spread of epidemic diseases like cholera. But on closer inspection, nature-induced disasters also turn out not to be random, unforeseen incidents. Rather, they are the culmination of long-ranging processes that are closely connected to societal activities (if not immanently a *result* of societal activities), which will continue to develop after the event, with restoration work and the appearance of coping strategies (cf. Plate and Merz; Alexander). For instance, the evolution of flu viruses within wild birds as host animals is a natural—and normal—process. Whether this "avian flu" transforms into a threat for human beings and develops into a pandemic is a question of political decisions and the result of societal practices rather than of natural processes, as Mike Davis showed in his work on the societal side of the 'production' of the avian flu.

Of course, there *are* major natural events, processes that shift huge amounts of land mass (landslides) or snow (avalanches), or that produce heavy eruptions or shocks to the earth's surface (volcanoes or earthquakes). But they need the agency of humans in order to turn into catastrophes or disasters.[1] Two of the earthquakes which occurred in 2010 are comparable in terms of magnitude (7.1) but very dif-

---

[1] This is, admittedly, a rather anthropocentric perspective. But in line with the arguments of Ludwig Wittgenstein, George Spencer Brown, Edmund Husserl, Heinz von Foerster, Ernst von Glasersfeld, and many others, any observation needs an observer: the observer has a certain perspective from which she or he observes. Logically, in the case of humans, this is an anthropocentric perspective.

ferent in terms of their impact: the number of casualties in the wake of the earthquake in Haiti on January 12 is estimated at 250,000 to 300,000 people, while following the quake in New Zealand on September 3, there were no deaths and only a
few injuries (of course, this is only true for the first tremor—the strong aftershock
on February 11, 2011, which had a magnitude of 6.3, led to major destruction and
169 casualties). Whether a natural event turns out to be a catastrophe[2] for the affected collective obviously has something to do with its political and socioeconomic strength, as well as with its social practices—a bundle of aspects that is
in general expressed in the concept of vulnerability (cf. Bankoff et al.; Blaikie et
al.; McAnany and Yoffee).[3]

Furthermore, the evolution of the structure of modern societies is deeply linked
to the increase in complexity, contingency, and self-reflection, the inevitability of
decision-making in the condition of not-knowing, and the general openness of the
future, and therefore: with risk and insecurity (cf. Luhmann; Japp; Beck; for the
use of knowledge in hazard mitigation White et al.). However, it is in society
where the decisions on how to deal with the general insecurity of the future and the
unpredictability of events are made. It is *in* society where flood prevention, emergency plans for an earthquake or a tsunami, or land development plans for moun-
tainous areas threatened by avalanches are invented and practiced. It is *in* society
where dangers and threats to a collective are identified, are communicated as risks,
and measures for their mitigation, prevention, or management are arranged. In this
broader sense, risks are always social constructions—whether they are initiated by
natural forces or by societal practices.

This chapter examines the role of space in the construction of risk. This is not a
conventional approach, since risk tends to be studied in isolation (November
1523). Risk helps to transform spaces (cf. November), while at the same time, spa-

---

[2]   The United Nations "International Strategy for Disaster Reduction" (ISDR) defines a catastrophe as any event that leads to a disruption of the operational capability of a collective, causes
high numbers of casualties, huge material, economic, and ecological losses, and outruns the
capability of the affected collective to cope with the event by its own means.

[3]   But also regions that are assumed to be resilient, and therefore very capable of coping with catastrophes, can be heavily affected, as we have recently seen in Japan. The series of grave incidents—the strongest earthquake ever measured (9.0), followed by a 33 ft high tsunami washing more than three miles inland and destroying the cooling system of the nuclear reactor of
Fukushima—would have been judged as "impossible" or "unthinkable" in any risk management simulation game. But it supports the notion that it is the social practices as well as the
societal decisions that lead to a catastrophe after an extreme natural event. To build nuclear
power plants despite its geological and tectonic situation was a well-discussed decision in Japan, which met with broad acceptance in Japanese society. The possibilities of a meltdown in
the wake of an earthquake were agreed on as a 'residual risk.'

tial indexing helps to construct risk (cf. Egner and Pott). I will also focus on the role of nature in the construction of risk. Space and nature both seem to play a similar role within the process of the social construction of risk. This is brought into focus by means of a closer study of the practices of risk research itself (cf. Egner and Pott).[4]

Some Conceptual Thoughts: On Risk, Space, and Observation Theory

Danger, Risk, and Safety: How Danger Turns into Risk

For almost three decades, following Ulrich Beck's constitutive study *Risk Society* (1992, in German 1986), risk has developed into one of the main concepts that can be used to structure modern societies (cf. Lupton; Renn). From this perspective, societal conflicts appear to be no longer "problems of societal order," but rather "problems of risk with open outcome" (cf. Bonß 17). Furthermore, in the social sciences, risks are seen as products of social interaction and not as facts that have to be determined and described. To grasp the meaning of a thing, it might be helpful to seek the appropriate antonym. As an antonym to risk, *security* seems to be the most logical contender. In this respect it is no surprise that most risk researchers assume that their findings will produce security (or safety) for a community, a collective, or for society in general (but then: how safe is safe enough, as Fischhoff et al. pointedly ask?). By using security or safety as an antonym to risk, the decision as to whether something proves to be risky or secure has to be postponed to the future, because it may take some decades before the results emerge. Therefore—and in order to shift attention to the societal aspects of risks—the sociologist Niklas Luhmann, as well as others (e.g. Japp), suggest utilizing the difference between risk and danger. These two terms can be distinguished by the attribution of decision: danger simply exists, while risk is the result of decisions. For instance, there is always the *danger* of getting an infection, but since there are vaccines available which inoculate against e.g. swine flu or H1N1, this danger is transformed into a *risk,* because it is up to the individual to decide for or against inoculation—and therefore: whether one can be blamed for getting ill or not. However,

---

[4]   In this study we applied observation theory to the field of risk research (Egner and Pott). We were interested in the question of who addresses what, when, in what context and with what consequences as a risk.

the inoculation comes with a new risk: the risk of unwanted—and maybe unknown—side effects.

Using danger as antonym to risk shows that risk is a relative concept. What is
risky for a single person or a collective can be dangerous for others—depending on
the decisions involved. For example, for those who made the decision to build a
nuclear reactor in a certain region, it is a risk. For all those who were not involved
into the decision-making process, the nuclear reactor is a danger. Hence, the question of what is classified as a risk is highly contingent, depending either on the
practice of construction within a specific culture of risk, or else on the perspective:
for whom and in what context a certain phenomenon is addressed as a risk or rather
as a danger. I will clarify this with the first example in section 3 of this chapter.
The decisions of the *National Oceanic and Atmospheric Administration* (NOAA)
concerning the complex natural situation at the mouth of the Saco River, Maine,
produced 'risks' for the state and local authorities. At the same time the decisions
established a 'danger' for the homeowners of the small beach community of Camp
Ellis who have build their houses on the newly created land, not knowing that the
sandy grounds can easily be washed away.

Spatial Aspects in Risk and Risk Research

In modern societies, different agents deal with risk. Government agencies, insurance companies, mass media, city and regional developers, and risk researchers—
all identify risks, pronounce warnings, and proclaim or question securities. Most
discourses on risk employ spatial facets or terminology, since hardly any risk is
distributed equally across a geographical area. Therefore, risks are mapped, localized, and spatially indexed. However, the effects of the spatiality of risk have not
yet been closely studied (cf. Egner and Pott; November). Any border that is drawn
on a map to denote risky or safe areas, e.g. in hazard zone maps, as well as any
(media) discussion of the locality in which an unwanted event happened, contributes to the transformation of the area in question. People use hazard zone maps for
the planning of new architectural or agricultural developments—in those areas de-
noted as 'safe'—or attribute specific semantics to particular localities, such as
'dangerous', 'no-go-areas', 'hazardous', etc. In doing so, they contribute to the
construction of these spaces as risky. There is obviously a need to refer to space in
order to transform dangers into manageable risks. In the broad sense, spatial indexing serves to make risks tangible and visible. It can be assumed that indexing risks
spatially is a key element in measures of risk management, with its purpose to reduce risk and to generate security.

Science, and especially risk research, plays a constitutive role in this discourse
and in the spatialization of risk. Niklas Luhmann states that in modern societies,
dangers are transformed into risks to the degree that its members (members of so-

ciety) are able to make decisions about preventative action (cf. 1993: 79 f.). Taking this view, then, it is "no accident that the risk perspective has developed parallel to the growth in scientific specialization" (28). Thus, science, as the core functional system in which knowledge is produced for society, contributes to the discourse by producing more and more contingent (= risky) knowledge: "the more we know, the better we know what we do not know, and the more elaborate our risk awareness becomes," as Luhmann puts it (28). Risk research encourages self-amplification of the societal risk debate. That leads to a paradoxical situation for risk researchers: on the one hand they are hoping and trying to generate security by means of their studies, on the other hand they are working on the distribution of a risk perspective that identifies risk everywhere, insisting on the societal importance of their work, and pointing out the need for action. By assisting government agencies, insurance companies, and other actors in the risk debate, science plays a major role in the so-cietal process of risk construction. By becoming simultaneously co-agent, medium of definition, and authority for solutions, science opens itself new markets for sci-entification and self-legitimation (cf. Beck 2009: 155 ff.). One might state, some-what provocatively, that we live in a risky world that is becoming even riskier thanks to the efforts of risk research.

Nature as a Social Category

The question of the relationship between 'nature' and 'culture' has a long history in both the sciences and the humanities (cf. e.g. Soper; Merchant; Latour), which, for reasons of space, must be omitted here. But some constitutive considerations need to be mentioned (cf. Egner 2008: 21 ff.): 'nature' has a paradoxical double meaning. On the one hand we are part of nature, and hence nature is our reference for any somatic or extra-somatic circulation of matter and energy. On the other hand, our relation to nature is always conducted and mediated by cultural imagina-tions which are based on the specific form of society and its cultural norms (cf. Brand 9). Moreover, a further interpretation of the double meaning of nature stresses on the one hand the genesis of the word 'nature', which in its Latin origin is a derivate of *nasci* = being born: from this perspective, nature is the essence, the marrow of a thing. On the other side, 'nature' embraces all elements and things that came into being without human assistance and which are subject to the 'laws of na-ture'. Here, nature is the counterpart of a culture that is defined by its creation by humans. If we are to take these arguments seriously, we must come to the rather unsettling conclusion, that culture is a product of humans, who at the same time are part of nature. Thus, culture must be part of nature, and cultural evolution part of natural evolution.

A contradictory current of thought holds that nature is part of culture since there is no understanding of nature possible without a cultural concept of it (cf. e.g.

Sixel; Soper). This perspective has some foundation in biological studies: Humberto Maturana and Francisco Varela state that we do not perceive the world on the basis of information on the things 'out there'. Rather, the experience of anything out there is validated by our biological and mental structure (cf. Maturana and Varela 25 f.). The inseparability of a particular way of being and how the world appears to us leads to the insight that every act of knowing brings forth a world. Thus, we bring forth the world by perceiving it. Our access to it is limited by our biological structure and mediated by our perceptions and experiences. That does not mean that there is no such thing as the 'real' world out there. But 'reality' is "experienced reality," as Bernhard Glaeser put it (58). Reality only becomes real through our specific, biologically structured perception, as well as through our thoughts, feelings, and our interpretations. From this perspective, nature *is* culture.

Obviously, these two perspectives are in opposition to one another, and there is no unifying view in sight, even if there have been attempts to overcome this form of dual thinking (cf. e.g. Latour; Law and Hassard). However, the traditional dichotomy of nature and culture still seems to be quite useful in structuring our world, and fulfilling the needs of political as well as scientific policy. For instance, the second case study in section 3 will show that in the debate on climate change, 'nature' is understood ontologically and is used in a reified way to stress the necessity of action, ignoring the notion that the currently favored narration of the 'greenhouse' is as culturally and socially assembled as all the other narrations in the global discourse on climate change.

Observation Theory and Risk Research

The analysis and explanation of why, who, when, in which context, and with which consequences a certain phenomenon can be defined as 'risk' necessitates methods for meta-observation of social practices and communication. Observation theory in the sense of Heinz von Foerster and Niklas Luhmann understands observation in two ways. First-order observation refers to any observation one can make—mechanically or intentionally. Some 'thing' is picked out by an observer who distinguishes it from all other things, and simultaneously names it. This observed 'thing' can be anything—from material objects to phenomena, or ideas; it can be a table, a landscape or a social process, just to give a few examples. The observer distinguishes on the level of what is being observed. The named 'thing' can be used as the starting point for further observation and communication. However, the distinction that has been used to differentiate the object or phenomenon from all other possibilities cannot be seen on this level. To observe the distinction of the observation on the first order level, one has to switch to second-order observation. It takes the first-order observation as a starting point from which to observe the distinction that has built the base for the first-order observation, because it can take all the other possibilities into account. With second-order observation the observer distinguishes on the level of 'how' (how did the first-order distinction come about?).[5] Applying this methodological concept to risk and risk research allows a deeper insight into the social practices of the construction of risk.

Observing 'Natural Catastrophes'—Two Case Studies

The following two case studies will show the employment of 'space' and 'nature' in the risk construction process. Despite the hazards faced by people living along shorelines, coastal areas are favored places to live all over the world, including in the United States. The first case study shows how spatial indexing as part of the risk management measurements to cope with natural hazard-induced risk does not in fact generate security, but rather leads to greater risks and dangers for coastal residents. The second case study focuses on the climate change debate and will show how scientific reference to 'nature' is employed to generate a new form of geodeterminism that reduces the discourse to a few scientific explanations and excludes crucial social aspects of the phenomenon.

The Role of Spatial Indexing: Coastal Communities in the United States

Coastal areas, and the areas on the banks of rivers, are popular locations for living and for business enterprises throughout the world, dating from the very beginning of human settlement. However, for some decades the ocean shoreline, especially with its view of the sea, has attracted growing numbers of people to build their houses right next to it. A report of the *National Oceanic and Atmospheric Administration* (NOAA) covering the years 1980 to 2008 states that 53 percent of US population growth is occurring in coastal regions that account for only 17 percent of land area. The population pressure on the shorelines is high—bridges are built to get to barrier islands and jetties are constructed to maintain inlet navigation. Furthermore, construction of dykes, seawalls, groin field and breakwaters is needed in order to protect coastal settlements from storms. Generally, these constructions are built under the assumption that the situation of a shoreline is 'naturally' stable and

---

[5]   Second-order observation does not necessarily mean seeing more or 'better,' but it is an appropriate method to tackle different dimensions of observation and to analyze the underlying differentiations that lead to a specific proposition.

predictable. Kelley et al. assembled a series of case studies of US coastal communities which try to manage the growing demand for real estate with a view of the sea, ignoring the scientific knowledge of beach areas possessing highly complex and dynamic structures. Living on a shoreline generates a 'coupled human-natural system,' i.e. "an integrated system in which people interact with natural systems," as Liu et al. have defined it (317). The complexity of such systems include non-linear dynamics, thresholds for alternate states within the system, temporal delay (often called 'legacy effect') between an action and its impact on either the human or natural side of the coupled system, and surprises, since the effects of complex dynamic systems are not predictable (cf. Werner and McNamara; McNamara and Werner).

An impressive example for how the neglect of the dynamics of a complex regimen with coupled natural-social systems leads to an expensive and almost un-winnable battle against 'nature' is the small beach community Camp Ellis, Saco, Maine (cf. Kelley and Brothers).[6] The province of Saco was settled by the British in the early seventeenth century, and along with the settlement, some sawmills were established along the Saco River. During the nineteenth century, textile plants gave rise to an energy demand that was mainly met by coal imported along water-ways on the Saco River. Navigation through the tidal deltas at the mouth of the Saco River was considered a crucial problem for the import and export of textiles and coal. For that reason, the mouth of the Saco River was dredged in 1837, and jetties were constructed to facilitate commercial navigation. As a result, the tidal delta collapsed, leading to beach accretion that was followed by residential devel-opment. The temporary accretion of sand at Camp Ellis altered the perception of this location from an unstable and thus uninhabitable environment to a static and stable ground appropriate for residential settlement, even if the ongoing beach ero-sion led quite rapidly to the loss of homes. People literally built their houses on sand without realizing the ephemeral structure of the location (Figure 1).

---

[6] My thanks to Margreth Keiler, Department of Geography and Regional Studies, University of Vienna, for the hint to use this example.

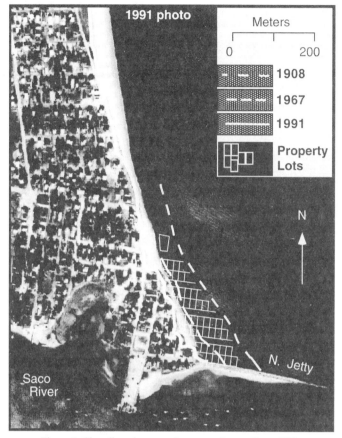

Figure 1: Shoreline change and property loss at Camp Ellis
between 1908 and 1991 (Kelley and Brothers 12)

However, the construction of jetties by the *US Army Corps of Engineers* (USA-COE) caused a change in the erosion and migration flows of sand. The erosion at Camp Ellis also led to a net longshore migration of sand to the north (instead of to the south, as the USACOE for decades assumed in their models) causing the closure of the Little River inlet, the growth of Pine Point spit about six miles north of Camp Ellis, and threatening to fill up the inlet of the Scarborough River as well. Thus, a jetty was placed at the tip of Pine Point to preclude the continued accretion, and Pine Point was developed for settlement in about 1940 (Figures 2).

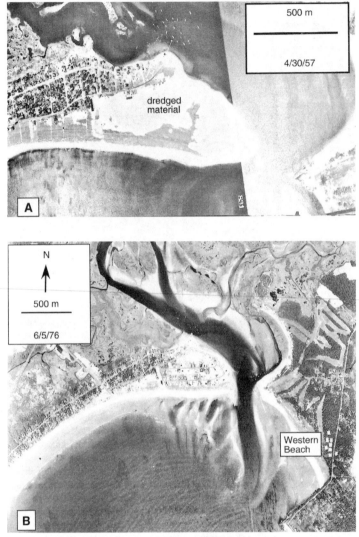

Figures 2: The air photos of Pine Point show the accretion of sand that had
already been dredged in 1957 (A) and the development of the settlement
on the bank in 1976 (B) (Kelley and Brothers 11)

Over several decades, the USACOE ignored the results of numerous studies that
showed the connection between beach erosion at Camp Ellis and beach accretion at
Pine Point. Recently, the USACOE conducted new research with detailed model-
ing of the tidal situation, the currents and the shoreline responses, which has led to
the proposal for the construction of new breakwaters seaward of Camp Ellis to
solve the problem of the ongoing erosion and continued threat to homeowners.

Based on the assumption that the situation at Camp Ellis and Pine Point is a complex coupled human-natural system, the geologists Kelley and Brothers judge the proposed solution very critically:

> Although the USACOE believes the breakwaters will "solve" the problems at Camp Ellis, in all likelihood, the overall system, people, and beach will each respond on their own. Property values would soar in the newly protected Camp Ellis; high-rise buildings and marinas are reasonable expectations because the added infrastructure (and the taxes paid by residents) would help defray the cost of necessary, perpetual beach replenishment. Just as Pine Point gained at Camp Ellis' loss earlier, Camp Ellis' gain now will likely come at the expense of their neighbors to the north. Wave refraction around the breakwaters could draw sand from the nearby beach into the protected area behind the breakwater, leading to erosion of properties that are in no danger today. It is conceivable that these homeowners will take action themselves some day to request their own breakwater. (17)

Even if it is a complex situation with many sources, Kelley and Brothers pinpoint "a principal figure in this tale of failed coastal management" (18): the USACOE. The proposed solution is, in a way, 'more of the same' in a series of 'risk management decisions' of the USACOE at Camp Ellis. Firstly, it authorized a project to improve the navigation at Saco River with minimum economic justification, thereby severely altering the sedimentation pathways in Saco Bay, and spending scant energy on investigating the spatial and temporal effects of the project. Secondly, the altering of the sedimentation pathways created new land at Camp Ellis. Despite the ephemeral structure of it, the USACOE designated this land for settlement development. Furthermore, the USACOE clung on to the navigation project after the exodus of the industry for which it was initiated in the first place. Thirdly, the USACOE ignored newer data about the currents and shoreline responses and almost dogmatically stuck to their conception of the net longshore currents to the south (Kelley and Brothers call it a "dogma" (18)) instead of realizing the net longshore sedimentation flow to the north—with all the effects on the river inlets there and accretion of sand at Pine Point. By not adapting their concepts in the light of the ongoing erosion at Camp Ellis and the accretion at Pine Point over the hundred years that passed following the construction of the northern jetty at Camp Ellis, the USACOE literally created spaces for more than a century, assuring the public that these areas were safe for building homes on. But unfortunately, the newly created spaces are highly risky ones for the residents of both Camp Ellis and Pine Point.[7]

Camp Ellis is just one example from the intensively inhabited coastline of the United States. Some communities seem to be a "model for maximizing coastal hazard vulnerability," as Pilkey and Neal ironically put it when talking about North Topsail Beach in North Carolina (73; for other examples cf. e.g. Rosen et al.; Wanless). Besides the problems which already exist here, it is conceivable that the struggle will intensify: the impending rise in ocean levels expected to occur as a result of ongoing global warming will provoke a myriad of 'shoreline responses' and will reinforce the critical situation for coastal homeowners. Trust in technical solutions as well as in agencies that pretend to generate 'security' for a community by means of risk management can only be justified on the basis of a deep understanding of the dynamics and complexity of coupled human-natural systems—and here we still have a long way to go.

'Nature' in the Discourse on Climate Change

Over the past few years the debate on 'climate change' has come to occupy a permanent place on the public agenda (on how the topic was placed on the public agenda, cf. Egner). Today, the importance of the issue is taken for granted—this is true at least of most Europeans and of the global scientific community. Here, referring to 'global warming' or 'climate change' no longer leads to a controversial scientific debate about the effects of global warming on the climate and on how the data can be interpreted. On the contrary, the references seem to be clear, distinct, and significant to everybody. Twenty years ago, this kind of common understanding did not exist. There was certainly a debate about climate change, but it consisted of very different 'narrations.'

The concept of 'narrations' is based on the assumption that humans, in essence, continue to be storytellers (*homo narrans*), despite the processes of modernization that involve more elaborate means of analysis, research, and understanding, as Walter Fisher points out (75). If this is true, then it is true also of scientists, who tell stories about their scientific findings, albeit using a more argumentative, denotative style (rather as I do in this paper) rather than a literary, aesthetically complex one. A narration is more than simply a story told by a narrator, and, significantly, it is not detached from social contexts, space, or time. Rather, narrations are schemes which build up systems of order that grant discourses meaning and coherence. In

---

[7]   The threatening situation in Camp Ellis and Pine Point can even be seen on the satellite images on Google Earth.

this way, narrations legitimate social and political practices, institutions, and ways of thinking, e.g. by justifying an inconvenient decision in the present by reference to the possibility of preventing a threat in the future (cf. Lyotard 49 f.; Viehöver 178).

Seen from this angle, the development of narrations about climate change offers some interesting insights. Between 1970 and 1995 there were at least six different narrations on the changing of the climate, as Willy Viehöver found in his analysis. As table 1 shows, they did not all predict a warming, in fact some projected a cooling of the atmosphere. The narration of a 'new ice age,' for instance, was based on evidence that aerosols cause an effect in the atmosphere called global dimming, which leads to a cooling effect (cf. Stanhill and Cohen; Wild et al.).[8] Interestingly enough, the same global dimming effect was feared by the peace movement as something which could result from nuclear bombings. This is combined in the narration of 'nuclear winter'. Both groups referred to the same effect in the climate but created rather different schemes to pursue their specific goals by giving the discourse a certain meaning and coherence: the one side in favor of nuclear energy, the other in favor of nuclear disarmament and against nuclear energy. Moreover, there is the small group of scientists that hold the view that the processes we refer to as climate change are a natural cyclic phenomenon resulting from the variability of sunspots. This narration is still out there, but no longer dominant. Over time, the differences between the various narrations have become more and more elaborate, but this did not lead to the coexistence of the different narrations, as has been observed in other scientific discourses.

---

[8]   The cooling effects are still measurable, but now they are understood as a factor that may have partially masked the effect of greenhouse gases on global warming. Moreover, the deliberate manipulation of this dimming effect is now being considered as a geo-engineering technique to reduce the impact of global warming.

narration category	green-house	new ice age	nuclear winter	climate paradise	cycles of sunspots	fiction of media and scientists
context	threat for human society, as well as flora and fauna, from anthropogenetically produced global warming	threat for human society from an anthropogenetically produced ice age	nuclear winter and atomic arms race between national super-powers	climate change as a chance for human society	climate change as cyclic, non-catastropic change	there is no climate change, it is merely the result of catastrophism of the media and science fiction
core problem	warming	cooling	cooling	warming	cyclic change	none
cause/ reason	CO_2 and other gases	aerosol, volcanoes, industry	cold war	technical mastery of nature	normal cyclic variability	none
strategies and goals	naturalizing and re-duction of climate change	pro nuclear energy	pro nuclear disarma-ment, no nuclear energy	optimism	fatalism	critique of media, re-jection of politicized scientists
protagonists	'heroes': scientists, institutions, environ-mental movement	University of Mary-land, NASA	peace movement	socialist countries (until 1990)	individual scientists, low media presence	MIT, US-Govern-ment (Bush)

Table 1: The different narrations in the discourse on climate change since 1974 (cf. Viehöver; Müller-Mahn); currently, only two of them are still subject to controversial opinion: the greenhouse narration and the climate skeptics

Today, the debate on climate change is dominated by just one narration: the 'greenhouse effect' narration, with its focus on an anthropogenic warming, measured by the increase of CO_2 in the atmosphere. There is hardly any doubt within the scientific community that there is a trend towards global warming which is triggered by human activity (IPCC). The only antagonist in this narration is the small but (in the United States) rather vocal group of climate skeptics, who see climate change as a piece of science fiction and the debate about it the result of catastrophe-oriented media coverage.

The interesting question is why the narration of the greenhouse effect was so successful. One could argue that it is a sign of scientific progress, that we know more and understand better than we did some years ago. But looking at the data of climate researchers, all that one can conclude is that there is a massive increase in

CO_2 in the atmosphere—we still don't know anything for sure about the effects of agents and their mutual reactions in this complex dynamic system. Thus, and quite correctly, the IPCC and other climate experts argue using probabilities and scenarios, instead of certitude and facts. By contrast, in the public debate, mass media as well as some climate scientists—mainly in their role as policy consultants—argue quite differently: climate change is constructed as the reaction of the overpowering force of nature with which humans are now confronted. Seen like this, climate change is the result of a disordered relationship between humans and nature, in which humans have ignored the general constraints of nature on human action, and have therefore created a situation in which climate change seems to be unpreventable. By the same logic, the IPCC, in its 2007 report, focuses on the adaptation to unavoidable climate changes as "responses to climate stimuli;" a phrase that reveals quite clearly a geodeterministic mode of thinking—a perspective we thought long overcome. Geodeterminism refers to a view that reduces complex socio-ecological interrelations to the aspects that can be detected by natural sciences. Moreover, geodeterminism tries to explain societal and cultural development by means of processes in the natural environment (cf. Blaut; Peet). Recently, some of these attempts have become quite popular, for instance Jared Diamond's effort to prove his thesis that "environment molds history" or David Landes' naïve explanation for the reasons for global inequality as the result of "nature's unfairness."

Within the very popular narration of the greenhouse effect, several geodeterministic arguments can be found:

- climate is constructed as part of a nature able to 'strike back' by means of floods, drought, storms, etc.; nature is understood ontologically and reified under the assumption that 'things are as they are,' and excluding contingency, as well as social and cultural impact;
- climate change is addressed as a massive constraint of future human action, potential adaptation is understood as a reaction to nature's stimuli;
- by addressing climate as the "heritage of mankind" (cf. Viehöver), climate is no longer understood as a neutral object, but rather as a sacred (but still ontological) force that we have to protect (and leave unchanged) for future generations. The 'heroes' of this protection are climate scientists, institutions like the IPCC, environmental groups, Al Gore, and (for a short period of time) the German chancellor Angela Merkel etc.;
- in protecting the sacred value of the climate system, it seems to be appropriate to focus on 'worst-case' scenarios by not indicating the probabilities, and to argue the certitude that all of the worst-case scenarios seem to be unavoidable;
- it also seems appropriate to select certain assumed causalities, while other possible causalities are ignored, and so once again to exclude contingency.

These aspects are an essential part of the greenhouse narration and they are successfully applied in public and political debate. Assuredly, this is not best scientific practice, but the goal of this politicized scientific communication is obvious: to emphasize the acuteness of action. Referring to nature as an ontological figure seems appropriate under the circumstances. But the reduction to such a geodeterministic view, with its employment of 'nature,' has proved rather risky. The 'climategate' in November 2009, as well as catastrophic failure of the conference in Copenhagen one month later, demonstrated the limits of this narration. Scientific findings that are presented with such explicit certainty just have to be true! Or, they have to be presented as 'mere' theories with the likely disadvantage of carrying much less weight in both the political and the public domain. As a result of the incidents in 2009, the credibility of climatology (and possibly even of science in general) is at stake, and a new, trustworthy climate policy, one which includes theories and uncertainties (for a first attempt cf. Prins et al.), as well as a new climate narration that gives the debate coherence and meaning—meaning which is not based on a notion as unstable as 'nature'—is desperately needed.

On the Role of Nature and Space in the Construction of Risk

Any determination of risk presumes observation, and thus an observer who distinguishes and names something as a risk. Thus, everything that is denominated as a risk in a society is contingent and the result of a processual and contextual dynamic in societal communication. The ongoing increase in the communication of risk indicates that the notion of 'risk' obviously produces a potent scheme of order for societal processes and phenomena that is productive both for individuals and whole segments of society.

The practice of indexing risk spatially is obviously also a very reasonable and well-regarded practice, since most of the communicated risks are linked to some spatial terminology. But why is that? Applying second-order observation shows that linking space to risk serves to make risks tangible and visible. Drawing a line or sketching out an area on a map gives advice about which areas are risky and which safe. Furthermore, spatial indexing also serves to conceal the uncertainty and the open outcome of phenomena. Indexing risk spatially can be seen as a process of reification of risk, with the feasible result that society can treat these risks as real and that the risks seem to be manageable. Thus, spatial indexing produces quasi-objective knowledge, used for generating social stability because of the supposed unambiguousness. Unfortunately, the assumed 'security' produced by these means includes some unwanted side effects: new risk.

The reification of risk is often linked to 'naturalization.' By referring to the 'nature' or to the 'natural state' of a risk, one can pretend that the addressed risk 'ex-

ists,' that it follows the 'laws of nature,' as the case study about the debate on climate change showed. By naturalizing risks, politics as well as society as a whole
'exculpates' itself of its burden of responsibility to address them, since nature as-
sumedly is 'just there' and follows its own rules.

All this leads to the following conclusion: because of their practices of spatialization and naturalization in the process of risk construction, risk researchers face a
dilemma: by aiming to contribute to the generation of societal security and safety
in their work, they in fact contribute to the production of new risks. However, it is
no solution whatsoever *not* to spatialize in risk management. We have to spatialize
in order to be capable of acting, to know who is at risk and where, so that institutions and organizations can address the issue and allocate their resources to allevi-
ate suffering. But a change in science seems to be inevitable: instead of representing their findings in the mode of experts, researchers would be well advised to
present the contingency and the uncertainties of the risk construction process together with their findings. This is in line with the understanding of Bechmann and
Stehr, who state that the central task of science is "not the announcement of assured knowledge but the management of uncertainty" (121).

Works Cited

Alexander, David. *Principles of Emergency Planning and Management*. Harpen-
den, UK: Terra Publications, 2002.
Bankoff, Greg, Georg Frerks, and Dorothea Hilhorst (eds.). *Mapping Vulnerability*.
London: Earthscan, 2004.
Bechmann, Gotthard and Nico Stehr. "Risikokommunikation und die Risiken der
Kommunikation wissenschaftlichen Wissens: Zum gesellschaftlichen Umgang
mit Nichtwissen." *GAIA* 9.2 (2000): 113-21.
Beck, Ulrich. *World at Risk*. Cambridge, UK: Polity Press, 2009.
———. *Risk Society. Towards a New Modernity*. London: Sage, 1992.
Blaikie, Piers et al. *At Risk: Natural Hazards, People's Vulnerability, and Disas-
ters*. 2nd ed. London: Routledge, 2003.
Blaut, James M. "Environmentalism and Eurocentrism." *Geographical Review*
89.3 (1999): 391-408.
Bonß, Wolfgang. *Vom Risiko: Unsicherheit und Ungewißheit in der Moderne*.
Hamburg: Hamburger Edition, 1995.
Brand, Karl-Werner. "Soziologie und Natur - Eine schwierige Beziehung. Zur Ein-
führung." *Soziologie und Natur: Theoretische Perspektiven*. Ed. Karl-Werner
Brand. Opladen: Leske + Budrich, 1998. 9-29.
Davis, Mike. *The Monster at Our Door: The Global Threat of Avian Flu*. New
York: The New Press, 2005.

Diamond, Jared. *Collapse: How Societies Choose to Fail or Succeed*. New York: Viking Adult, 2004.

Egner, Heike. *Gesellschaft, Mensch, Umwelt – beobachtet. Ein Beitrag zur Theorie der Geographie*. Stuttgart: Franz Steiner, 2008.

———. "Überraschender Zufall oder gelungene wissenschaftliche Kommunikation: Wie kam der Klimawandel in die aktuelle Debatte?" *GAIA* 16.4 (2007): 250-54.

Egner, Heike and Andreas Pott. "Geographische Risikoforschung beobachtet." *Geographische Risikoforschung: Zur Konstruktion verräumlichter Risiken und Sicherheiten*. Ed. Heike Egner and Andreas Pott. Stuttgart: Steiner, 2010. 231-39.

——— (eds.). *Geographische Risikoforschung: Zur Konstruktion verräumlichter Risiken und Sicherheiten*. Stuttgart: Steiner, 2010.

Fischhoff, Baruch et al. "How Safe is Safe Enough? A Psychometric Study of Attitudes Towards Technological Risks and Benefits." *The Earthscan Reader on Risk*. Ed. Ragnar E. Löfstedt and Åsa Boholm. London: Earthscan, 2009. 87-112.

Fisher, Walter R. "The Narrative Paradigm: In the Beginning." *Journal of Communication* 35 (1985): 74-89.

Foerster, Heinz von. *Observing Systems*. 2nd ed. Seaside, CA: Intersystems Publications, 1984.

Glaeser, Bernhard. "Natur in der Krise? Ein kulturelles Mißverständnis." *Humanökologie und Kulturökologie: Grundlagen, Ansätze, Praxis*. Ed. Bernhard Glaeser and Parto Teherani-Krönner. Opladen: Westdeutscher Verlag, 1992. 49-70.

IPCC, Intergovernmental Panel on Climate Change. *Climate Change 2007: Impacts, Adaptation and Vulnerability. Contribution of Working Group II to the Fourth Assessment Report of the Intergovernmental Panel on Climate Change*. Ed. M.L. Parry, O.F. Canziani, J.P. Palutikof, P.J. van der Linden, and C.E. Hanson. Cambridge, UK: Cambridge University Press, 2007.

———. *Climate Change 2007. The Physical Science Basis. Contribution of Working Group I to the Fourth Assessment Report of the Intergovernmental Panel on Climate Change*. Ed. S. Solomon, D. Qin, M. Manning, Z. Chen, M. Marquis, K.B. Averyt, M. Tignor, and H.L. Miller. Cambridge, UK: Cambridge University Press, 2007.

Japp, Klaus Peter. "Systems, Theory, and Risk." *Social Theory of Risk and Uncertainty*. Ed. Jens O. Zinn. Oxford: Blackwell, 2008. 75-106.

———. *Soziologische Risikotheorie: Funktionale Differenzierung, Politisierung und Reflexion*. Weinheim: Juventa Verlag, 1996.

Kelley, Joseph T. and Laura L. Brothers. "Camp Ellis, Maine: A small beach community with a big problem...its jetty." *America's Most Vulnerable Coastal*

Communities. Ed. Joseph T. Kelley et al. Boulder: Geological Society of America, 2009. 1-20.

Kelley, Joseph T., Orrin H. Pilkey, and J. Andrew G. Cooper (eds.). *America's Most Vulnerable Coastal Communities*. Boulder: Geological Society of America, 2009.

Landes, David S. *The Wealth and Poverty of Nations: Why Some Are So Rich and Some So Poor*. New York: Norton, 1999.

Latour, Bruno. *Das Parlament der Dinge: Naturpolitik*. Frankfurt am Main: Suhrkamp, 2001.

———. *We Have Never Been Modern*. Cambridge, MA: Harvard University Press, 1993.

Law, John and John Hassard (eds.). *Actor Network Theory and After*. Oxford: Blackwell, 1999.

Liu, Jianguo et al. "Complexity of Coupled Human-Natural Systems." *Science* 317 (2007): 513-16.

Luhmann, Niklas. *Risk: A Sociological Theory*. Berlin: de Gruyter, 1993.

———. *Beobachtungen der Moderne*. Opladen: Westdeutscher Verlag, 1992.

Lupton, Deborah. *Risk*. Milton Park: Routledge, 1999.

Lyotard, Jean-François. "Randbemerkungen zu den Erzählungen." *Postmoderne und Dekonstruktion: Texte französischer Philosophen der Gegenwart*. Ed. Peter Engelmann. Stuttgart: Reclam, 1990. 49-53.

Maturana, Humberto R. and Francisco J. Varela. *Tree of Knowledge: Biological Roots of Human Understanding*. Boston: Shambala, 1998.

McAnany, Patricia A. and Norman Yoffee (eds.). *Questioning Collapse: Human Resilience, Ecological Vulnerability, and the Aftermath of Empire*. Cambridge: Cambridge University Press, 2009.

McNamara, D. E. and B. T. Werner. "Coupled barrier island-resort model: 1. Emergent instabilities induced by strong human-landscape interactions." *Journal of Geophysical Research* 113 (2008). F01016, doi:10.1029/2007JF000840.

———. "Coupled barrier island-resort model: 2. Tests and predictions along Ocean City and Assateague Island National Seeshore, Maryland." *Journal of Geophysical Research* 113 (2008). F01017, doi:10.1029/2007JF000841.

Merchant, Carolyn. *The Death of Nature: Women, Ecology, and the Scientific Revolution*. San Francisco: Harper Collins, 1990.

Müller-Mahn, Detlef. "Beobachtungen zum Klimadiskurs: Neues Weltrisiko oder alter Geodeterminismus?" *Geographische Risikoforschung. Zur Konstruktion verräumlichter Risiken und Sicherheiten*. Ed. Heike Egner and Andreas Pott. Stuttgart: Steiner, 2010. 95-113.

NOAA, National Oceanic and Atmospheric Administiation. "Population trends along the costal United States: 1980-2008." 2004. Accessed 15 October 2010. <http://oceanservice.noaa.gov/programs/mb/supp_cstl_population>.

November, Valérie. "Spatiality of Risk." *Environment and Planning A* 40 (2008): 1523-27.

Peet, Richard. "The Social Origins of Environmental Determinism." *Association of American Geographers* 75.3 (1985): 309-33.

Pilkey, Orrin H. and William J. Neal. "North Topsail Beach, North Carolina: A model for maximizing coastal hazard vulnerability." *America's Most Vulnerable Coastal Communities*. Ed. Joseph T. Kelley et al. Boulder: Geological Society of America, 2009. 73-90.

Plate, Erich J. and Bruno Merz (eds.). *Naturkatastrophen: Ursachen, Auswirkungen, Vorsorge*. Stuttgart: Schweizerbart, 2001.

Prins, Gwyn et al. *The Hartwell Paper: A New Direction for Climate Policy After the Crash 2009*. Oxford: Institute for Science, Innovation and Society, University of Oxford, 2010.

Renn, Ortwin. "Concepts of Risk: Part I; An Interdisciplinary Review." *GAIA* 17.1 (2008): 50-66.

Rosen, Peter S., Duncan M. FitzGerald, and Ilya V. Buynevich. "Balancing Natural Processs and Competing Uses on a Transgressive Barrier, Duxbury Beach, Massachusetts." *America's Most Vulnerable Coastal Communities*. Ed. Joseph T. Kelley et al. Boulder: Geological Society of America, 2009. 21-32.

Sixel, Friedrich W. *Die Natur in unserer Kultur. Eine Studie in der Anthropologie und der Soziologie des Wissens*. Würzburg: Königshausen und Neumann, 2003.

Soper, Kate. *What is Nature? Culture, Politics and the Non-Human*. Oxford: Blackwell, 1995.

Stanhill, Gerald and Shabtai Cohen. "Global Dimming: A Review of the Evidence for a Widespread and Significant Reduction in Global Radiation with Discussion of its Probable Causes and Possible Agricultural Consequences." *Agricultural and Forest Meteorology* 107 (2001): 255-78. doi:10.1016/S0168-1923(00)00241-0.

Viehöver, Willy. "Die Wissenschaft und die Wiederverzauberung des sublunaren Raumes: Der Klimadiskurs im Licht der narrativen Diskursanalyse." *Handbuch, Sozialwissenschaftliche Diskursanalyse. Band 2: Forschungspraxis*. Ed. Reiner Keller et al. Opladen: Leske + Budrich, 2003. 233-69.

———. "Diskurse als Narrationen." *Handbuch Sozialwissenschaftliche Diskursanalyse. Band 1: Theorien und Methoden*. Ed. Reiner Keller et al. Opladen: Leske + Budrich, 2001. 177-206.

Wanless, Harold R. "A History of Poor Economic and Environmental Renourishment Decisions in Broward County, Florida." *America's Most Vulnerable Coastal Communities*. Ed. Joseph T. Kelley et al. Boulder: Geological Society of America, 2009. 111-20.

Werner, B. T. and D. E. McNamara. "Dynamics of Coupled Human-landscape Systems." *Geomorphology* 91 (2007): 393-407.

White, Gilbert F., Robert W. Kates, and Ian Burton. "Knowing Better and Losing Even More: the Use of Knowledge in Hazards Management." *Environmental Hazards* 3 (2001): 81-92.

Wild, Martin, Atsumu Ohmura, and Knut Makowski. "Impact of Global Dimming and Brightening on Global Warming." *Geophysical Research Letters* 34 (2007). L04702, doi:10.1029/2006GL028031.

Buffalo Commons:
The Past, Present, and Future of an Idea

Andrew C. Isenberg

In 1987, a pair of geographers from New Jersey, alarmed by increasing soil erosion and a declining population in the North American Great Plains, proposed reintroducing bison to a sizeable portion of the region and designating the area a "Buffalo Commons" (Popper and Popper 1987: 12). The proposal was polarizing: many environmentalists saw the idea as one that could heal a region that had suffered not only the near extinction of the bison in the late nineteenth century, but also the infamous Dust Bowl of the 1930s. Rural residents of the Great Plains, by contrast, were largely—and vociferously—opposed to what they see as a proposal by distant metropolitan intellectuals who understand neither rural culture nor the Great Plains environment.

Whatever one thinks of the merits of the Buffalo Commons idea, it certainly represents environmental restoration on a grand scale: Frank and Deborah Popper, the geographers who first proposed the idea, envisioned a region "returned to its original pre-white state" as "the settlers found in the nineteenth century" (1987: 12-18). Despite their all-encompassing, un-self-consciously frontierish rhetoric, it is unlikely that the Poppers thought that a complete return to a "pre-white state," when as many as thirty million bison inhabited the grasslands, would be possible throughout the Great Plains. Rather, what they and other proponents of the Buffalo Commons idea had in mind was reconstructing a simulacrum of the historic Great Plains in part of the grasslands. They imagined the historic grasslands in idyllic terms. Before the arrival of Euroamerican settlers, they presumed, the Great Plains had abounded with so many bison that natives lived in comfortable and sustainable harmony with them (Callenbach 34). Yet environmental history suggests a more complex reality. Recent work in grassland ecology suggests that the North American grasslands are a dynamic, highly unpredictable environment in which the historic bison population probably fluctuated considerably (Dillehay 180; Caughley 189; Forsythe and Caley 297; Petrie 6). In this changeable biome, if the natives' harvest of the bison was sustainable, it was only by the barest of margins (Isenberg 2000: 92). These complexities notwithstanding, advocates of a Buffalo Commons are sanguine about the reintroduction of the bison. Yet previous efforts to reintroduce bison to the Great Plains a century ago bore decidedly mixed results (Isenberg 1997: 180). In short, a Buffalo Commons—or any "green culture" in the Great Plains, for that matter—must take into account the twin dynamisms of history and ecology.

When the Poppers first proposed the Buffalo Commons idea, they saw the Great Plains as broken and wasted. Ranching and farming, they argued, had led to extensive soil erosion in the plains. The declining productivity of the land had led, in turn, to out-migration. For the Poppers, the chief event for understanding the plight of the Great Plains was the Dust Bowl of the 1930s. That ecological disaster followed the sudden expansion of wheat cultivation in the southern Great Plains in the first decades of the twentieth century. Farmers broke 32 million acres (13 million hectares) of sod in the grasslands between 1909 and 1929. Plowing up drought-resistant native grasses and sowing wheat produced (for a few seasons) record yields. One environmental historian has called this phenomenon "ecosystem harvest": a burst of productivity from nutrient-rich soils never before cultivated (Zarrilli 570). But when a decade-long drought began in 1932, the wheat shriveled and the exposed soil blew away; dust storms blew Great Plains soil across the continent; in 1934, as the environmental historian Donald Worster has evocatively narrated, the dirt fell like snow in Chicago, New York, Boston, and Atlanta. Dust Bowl refugees fled the region (Worster 13-19, 53). For the Poppers, the Dust Bowl was only the most notable instance of the century-long failure of American settlement of the Great Plains—what they called "the largest, longest-running agricultural and environmental miscalculation in American history." The Great Plains was "turning into an utter wasteland, an American Empty Quarter." The solution, they argued, was to reverse the flow of land into private hands, to "deprivatize"—an artful term for nationalizing private property— the Great Plains and create a Buffalo Commons, "the world's largest historic preservation project, the ultimate national park" (Popper and Popper 1987: 12-18).

Plains residents reacted to the Buffalo Commons proposal with disdain. The fact that the Poppers were a pair of urbanites from New Jersey, the most crowded state in the United States (its population density is equivalent to that of the Netherlands) and a place about as far from the Great Plains as one can be in the United States, both geographically and culturally, only made plains residents' reaction to the proposal more dismissive. "Don't try to come in and use our land for common property for people from New Jersey and California," one Nebraska farmer told the Poppers (Matthews 54). Though residents of the Great Plains conceded that the rural population of the region was declining, many resisted the idea that the solution to that problem was simply to accelerate the decline. Their reaction mirrored that of "conservation refugees" in the developing world—relatively powerless farmers whom states have displaced to create designated wilderness areas (Dowd 6). One can certainly understand the reaction: the Poppers' 1987 essay was meant to be provocative. Their cultural distance from the grasslands was evident in their proposal, which uncritically touted the advantages of a Buffalo Commons for urban eco-tourists. At the time they wrote their essay, they lacked much personal familiarity with the region. On a speaking tour after the publication of the essay, Frank

Popper alighted in Cheyenne County, Kansas, and declared the place "exotic," according to a journalist (Matthews 48). What many plains residents regard as the region's advantages—that it is relatively unpopulated and rural—the Poppers regarded with evident horror: "A dusty town," they wrote, "with a single gas station, store, and house is sometimes 50 unpaved miles from its nearest neighbor, another three-building settlement amid the sagebrush" (Popper and Popper 1987: 12-13). This place, they implied, was fit for bison but not for people.

As the Poppers noted, such a sparse human population was the result of several decades of rural out-migration—a trend they regarded as irreversible. "A few of the more urban areas may pull out of their decline," they wrote, "but the small towns and surrounding countryside will empty, wither, and die." The United States Census Bureau data bears out the Poppers' foresight. Between 1990 and 2007, the Great Plains—376 counties spread across ten states—actually increased in population, from 7.6 million in 1990 to 9.9 million in 2007. But as has been the case since 1950, population growth was concentrated in one-third of Great Plains counties on the margins of the region, notably a belt of metropolitan counties in Colorado containing the cities of Boulder and Denver, and a similar belt of metropolitan counties in Texas containing Waco, Austin, and San Antonio. The rural two-thirds of Great Plains—244 of its 376 counties—lost population.

To put the region's depopulation in perspective, consider that over a century ago, the Census Bureau, having analyzed its 1890 data, declared that a discernable frontier line no longer existed in the United States. After a century of settlement in the West, most counties in the nation had achieved a population density greater than two persons per square mile (roughly two persons per 250 hectares). Reaching that threshold was the result of federal policies designed to privatize public land in the Great Plains and encourage settlement—the most well-known policy initiative was the Homestead Act of 1862, which offered settlers 160 acres (65 hectares) at no cost. Yet rural depopulation in the last half-century has caused the frontier to reappear: three-quarters of Great Plains counties have population densities of less than two persons per square mile.

In 1893, the historian Frederick Jackson Turner seized on the 1890 census data to argue in one of the most important and enduring historical essays ever published, "The Significance of the Frontier in American History," that the settlement of the frontier—what he called "the transformation of wilderness to civilization"—was the dominating characteristic of American history up until 1890; "the existence of an area of free land, its continuous recession, and the advance of American settlement westward explain American development." Frontier settlement made Americans free (because tyranny thrived only in the crowded Old World, according to Turner) and prosperous (because of the abundant resources of the land). But with the closing of the frontier, that chapter of American history had ended (F. Turner 199). Turner's essay was always more rhetorical than real. American historians have demolished his generalizations and exposed his fallacies: "free land" was inhabited by natives; the Homestead Act (which came relatively late in the settlement process) notwithstanding, most land was expensive and rural settlement was characterized by squatting and tenancy; most Americans in the nineteenth century migrated not from crowded cities to open land but from rural to urban areas (Gates 60; Shannon 31). Nonetheless, Turner's celebratory, progressive narrative of American history remains stubbornly durable in popular American history.

Understanding Turner is essential to understanding both the Poppers' proposal for a Buffalo Commons and their opponents in the Great Plains. The Poppers' proposal was an effort to restore the wilderness that, as Turner wrote, was the wellspring of American values and identity. As in existing national parks, similarly established in the late nineteenth and early twentieth century to preserve a remnant of frontier wildness, Americans would not settle in the wilderness of the Buffalo Commons, but visit it and feel its salutary effects. Great Plains farmers and ranchers, by contrast, regard themselves as the happy result of Turner's settlement narrative—independent, prosperous, and practical.

Since 1987, the common Turnerian roots of both the Poppers and their opponents have helped the two sides move closer together. Following several years of speaking engagements in the plains—encountering resistance and fierce criticism from some of the remaining residents of the rural plains whom the Poppers had slated for extinction—the Poppers retreated from their call for the deprivatization of the region. In retrospect, the Poppers had always been uneasy about this part of their proposal, which is why even in 1987 they had substituted the term "deprivatize" for the more politically-charged term "nationalize." Their retreat from deprivatization was also, in part, recognition of the national conservative political ascendancy that had sharply attenuated the federal government's willingness and ability to add new lands to its system of national parks and wilderness areas (J. Turner 244). By 1999, the Poppers had redefined their call for a Buffalo Commons as a "metaphor" for rethinking land use in the Great Plains. The term was, they wrote, nothing more than a "literary device," an "umbrella phrase for a large-scale, long-term restoration project." Their new "soft-edged" Buffalo Commons abandoned their earlier strict call for nationalization of private property to protect a federal bison preserve. They welcomed efforts by ranchers and Indian reservations to create bison ranges (Popper and Popper 1999). Their call for a Buffalo Commons became, like Turner's frontier thesis, an exercise in rhetoric that many people could support while maintaining quite different interpretations of it. They kept the idea vague enough to allow such different interpretations.

The Poppers were not the first to call for a Buffalo Commons in the Great Plains. In 1905, a group of conservationists including President Theodore Roosevelt founded the American Bison Society, which over the next decade established a half-dozen preserves for the bison, including the National Bison Range in Montana. Still earlier, in 1832—forty years before the United States government created the world's first national park in Yellowstone—the itinerant artist George Catlin, while touring the Missouri River Valley, reflected that the bison was "so rapidly wasting from the world, that its species must soon be extinguished." Catlin recommended that the federal government create "a *nation's Park*" in the grasslands. He imagined that both the bison and the natives who hunted them "might in future be seen (by some great protecting policy of government) preserved in their pristine beauty and wildness, in a magnificent park" (Catlin 294-95). And well before Catlin, beginning in the mid-eighteenth century, native equestrian nomads including the Lakota, Cheyenne, and Crow subsisted on the bison, managing a "buffalo commons" that encompassed most of the Great Plains (Isenberg 2000: 8). What insights does the environmental history of the bison—which might be thought of as past experiences with versions of the Buffalo Commons—provide us?

The foremost insight that environmental history offers is the understanding that the Great Plains environment is dynamic. The trophic dynamics of the Great Plains constitute what one may call a *solar economy* (Pfister). Grasses absorb the energy of the sun and, through photosynthesis, transform part of that energy into carbohydrates. Bison transform some of those carbohydrates into protein. Predators (wolves and humans) hunt grazing animals and consume their protein. The introduction of domesticated grasses such as wheat, rye, and alfalfa and grazing animals such as cattle have altered and domesticated the floral and faunal constituency of the Great Plains without changing its basic trophic structure. Some energy from the sun is lost as it moves from one trophic level to another—from grasses to herbivores to carnivores—so at each trophic level above grasses there exists less chemical energy in the form of food. A grassland ecosystem is thus pyramidal in structure: extensive grasses support a sizeable but smaller biomass of grazers, which in turn support a smaller number of predators. Nonetheless, because the energy of the sun feeds the grasses at the base of the pyramid, theoretically the system is indefinitely renewable, so long as the population of grazing animals does not outstrip the supply of forage, and predators (human or otherwise) confine their harvest of grazing animals to a sustainable level (Goudie 322).

Yet the Great Plains has been characterized not by renewable stability, but by unpredictable change. The near extinction of the bison in the early 1880s was followed in short order by the collapse of the free-range cattle industry later in the decade and the Dust Bowl of the 1930s. These human failures reflect the dynamism of the Great Plains environment. Located on the leeward side of the Rocky Mountains, the plains lie in the mountains' "rain shadow" that inhibits the flow of eastward-moving, moisture-bearing air currents. Average annual precipitation in the Great Plains is 20 inches (500 mm) or less; grasses are the dominant vegetation because their above-ground structures are minimal (and thus require minimal moisture to maintain) while their dense roots close to the surface take advantage of available soil moisture. Precipitation in grasslands is typically not only scarce but unpredictable. In the Great Plains, a few successive rainy years will as likely as not be followed by a period of drought. Intervals between droughts can be anywhere from fifteen to thirty years; the droughts themselves can last anywhere from a year to ten years or longer. The infamous drought in North America in the 1930s that helped cause the Dust Bowl is perhaps the most well known, but it was by no means an anomaly in the Great Plains. Dendrochronological studies dating from the mid-sixteenth century show that precipitation in the Great Plains has been extremely variable; the region has been characterized by many periods of low precipitation lasting ten years or longer (Isenberg 2000: 17-18). "Herein lies the major problem of steppe regions," wrote the geographers Robert Gabler, Robert Sager, Sheila Brazer, and Daniel Wise. They "seem like better-watered deserts at one time and like slightly subhumid versions of their humid climate neighbors at another" (Gabler et al. 210).

The marginal, dynamic grassland environment imposed limits on the historical population of the bison. Euroamerican observers in the nineteenth century estimated that the total bison population in North America might be as high as 75 or 100 million; for most of the twentieth century, ecologists and historians believed that the historic bison population probably numbered no fewer than 60 million (Seton 654-56; Roe 489-520). In more recent years, however, more sober estimates of the historic bison population, based on realistic assessments of the carrying capacity of the grasslands, have led scholars to the conclusion that between 24 and 30 million bison grazed in the Great Plains in the late eighteenth century (McHugh 16-17; Flores 470-71; Isenberg 2000: 28-30).

Whatever the upward limit of the bison population might have been, in the dynamic, unpredictable Great Plains environment, the bison population was rarely stable. Bison, like other ungulates, are usually at disequilibrium with their forage: their populations are apt to irrupt, placing too much pressure on the range, so that population irruptions are followed by abrupt declines (Melville 6-7). Drought periodically caused significant declines in range carrying capacity and consequently in the bison population (Isenberg 2000: 83-84). In the southern Great Plains in places where prehistoric American hunters drove bison to their deaths there were two long periods—from 6000 to 2500 BC and from 500 to 1300 AD—when bison were absent from the kill site remains, indicating steep declines in the bison population (Dillehay 180-96).

There were enough bison in the Great Plains to draw the attention of native hunters. Yet when Europeans first came to North America, such groups as the Arapahoes, Assiniboines, Atsinas, Blackfeet, Cheyenne, Comanche, Crow, Kiowas, and

Lakota subsisted on the fringes of the grasslands. They had no horses, and while they traveled seasonally to the Great Plains to hunt bison on foot, they relied primarily on hunting and gathering (and in some cases, farming) in the regions outside the grasslands. It was, ironically, what the historian Alfred Crosby has called European "ecological imperialism" that created the plains nomadic societies (xiv). Spanish colonists introduced horses to North America in the sixteenth century (or, to be more precise, they reintroduced horses. Equines originated in the Americas, crossed westward over the Bering Strait land bridge to Asia, and eventually became extinct in the Western Hemisphere). In reaction to the Europeans' (re-) introduction of the horse (which facilitated bison hunting) and Old World diseases such as smallpox (which discouraged agricultural settlements, where diseases thrived among dense populations), the Lakota, Cheyenne, and others reinvented themselves as equestrian nomads on the Great Plains.

The stereotype of the plains nomads imagines them living in harmony with the bison: hunting only when necessary and wasting no parts of their kills—this is the romantic image many proponents of the Buffalo Commons idea invoke to support their proposal. The natives' culture was, in the minds of many modern-day environmentalists, the archetype of a green culture (Hughes 7, 28). Yet nomadic bison hunting was not a tested, time-honored resource strategy. It was an experiment, an eighteenth-century response to the ecological changes created by the European conquest of America. The newly created nomadic societies were unlike most foraging groups, which derived 80 percent or more of their subsistence from gathering. A reliance on gathering, many ecological anthropologists have argued, is not only a stable resource strategy, but requires relatively little labor (Lee 30-43). Yet the Great Plains nomads relied primarily on hunting large mammals, a strategy that is far less reliable owing to the unpredictability of game movements and population. The strategy can also be exceptionally destructive to game resources. The reindeer, caribou, sea otter, seal, whale, and musk ox hunters of the North American and the Eurasian Arctic, for instance, are known to have practiced wasteful hunting techniques, such as selecting cows and calves and killing large numbers of animals during breeding seasons, when the animals conveniently assembled in large numbers (Ingold 69-75; Krupnik 233-39). The plains nomads also selected cows and killed large numbers of bison during the breeding season.

There were social as well as environmental costs to nomadic bison hunting. The nomads' constant movement in search of the bison atomized their societies. The nomadic societies could assemble in large numbers for communal hunts only during the summer, when 75 percent of regional precipitation falls. During the summer rainy season, grasses thickened and the bison assembled for the mating season. At other times of the year, the nomadic societies splintered into small foraging bands (Isenberg 2000: 43).

Even divided into bands, the nomads' subsistence was uncertain and famine was familiar. In these circumstances, most groups discouraged overhunting in order to minimize the risk of famine. The average nomad probably consumed six or seven bison each year for sustenance. The nomads likely killed a further number of bison for intertribal trade. The estimated 60,000 nomads in the plains in the nineteenth century thus probably killed 450,000 bison every year—almost all of them two- to five-year-old cows selected for their tasty meat and pliable hides (Isenberg 2000: 83; Flores 481-82). Such a harvest was seemingly sustainable. Yet ecological factors—wolves, drought, grassfires, droving, blizzards, competition from other grazers—unpredictably influenced the bison's population and could render the nomads' restrained use of the bison unsustainable. In some years, ecological factors could have combined to exceed the natural increase of the species (Isenberg 2000: 84).

The natives' nomadic experiment—adapting to the introduction of the horse and constructing societies built around the hunting of the bison—is instructive. Adapting to the solar economy of the Great Plains is a challenge—indeed, it is an ongoing, never-ending challenge. The grassland environment is dynamic; a sustainable harvest of bison is not a fixed sum, year in and year out, but fluctuates in response to unpredictable environmental pressures. Perhaps, had Euroamericans not overrun the Great Plains after the mid-nineteenth century and replaced bison with domesticated cattle, the Great Plains nomads might have demonstrated the flexibility to sustain their nomadic experiment. Yet had Europeans not come to North America, neither would have the horses essential to the nomadic societies. Human societies cannot arrest economic, cultural, and environmental change. The history of the Great Plains nomads suggests that as human societies seek sustainability—as they seek to create a green culture—they must continually adjust to social and ecological change.

If a green culture must be adaptive, at the same time we must bear in mind that cultures are complex and often contradictory. No culture in the Great Plains has ever single-mindedly pursued sustainability. All have imposed on the environment—and on the bison in particular—other cultural demands as well. Consider the nineteenth-century Lakota, who, like other nomadic societies in the Great Plains whose subsistence depended on the bison, maintained customs that strongly discouraged the overhunting of their primary resource. Yet at the same time, during the summer, when the bison assembled in huge herds for the mating season, the Lakota like other nomads often splurged on wasteful feasts of fresh meat, deliberately squandering bison, taking only the best parts of the fattest cows or killing more bison than they required at a time. During her captivity among the Lakota in 1864, Fanny Kelly observed her captors' wasteful use of the bison during the summer. "The Indians often, for the mere sport, make an onslaught, killing great numbers of them, and having a plentiful feast of 'ta-tonka,' as they call buffalo meat," she wrote. "Each man selects part of the animal he has killed that best suits his own taste, and leaves the rest to decay or be eaten by wolves, thus wasting their own game" (Kelly 76).

How could the nomads, who so depended upon the herds, be so wasteful? The hunts that Kelly described were special events. The summer feasts served an important social purpose. Divided into foraging bands during the lean winter and spring months, the nomads often endured famine. The nomads came together in large groups during the summer, however, mirroring the pattern of the bison, who assembled in the summer for the mating season. When they congregated together in the summer, the nomads affirmed their social solidarity by squandering the resources of the large summer herds. The nomads distributed the proceeds of the summer hunt to all (Isenberg 2000: 90).

Similarly, those who sought to preserve the bison from extinction a century ago—a handful of American and Canadian wildlife advocates—likewise exploited the bison to serve cultural ends. The United States government stocked Yellowstone Park with bison in 1902—to the delight of the Northern Pacific Railroad, a company that had lobbied for the creation of Yellowstone Park in 1872 as a tourist destination for their passengers. Between 1905 and 1914, under pressure from the American Bison Society, an organization of well-connected and well-heeled sport hunters and wildlife advocates, the United States government founded bison preserves in Oklahoma, Montana, South Dakota, and Nebraska. The Canadian government established bison preserves in Alberta in 1907 and 1922. These preservationists confined the bison to small reservations suited to tourism. As a result, by the first decades of the twentieth century the bison had become a largely domesticated species maintained only by the constant intervention of human keepers.

In many respects, this limited result was exactly what the early twentieth-century bison preservationists intended. Although the preservationists regretted the wasteful slaughter of the bison by Anglo hide hunters in the 1870s—a spasm of hunting that delivered the *coup de grâce* to the bison—their primary animus was their unease with modern, urban, industrial society. For preservationists, the bison was a mythic symbol of untamed nature, the frontier, and, most importantly, masculinity—since they considered the conquest of the frontier a male endeavor. Like Frederick Jackson Turner, they mourned the passing of a frontier they considered essential to American identity.

In 1902, as part of his effort to preserve that identity, President Theodore Roosevelt—an ardent conservationist who would become the honorary president of the American Bison Society when it was founded three years later—appointed Charles Jesse "Buffalo" Jones—a former "Wild West" showman much like "Buffalo Bill" Cody—to a newly-created position: Game Warden of Yellowstone. Jones fenced in part of Mammoth Valley and stocked it with twenty-one bison that he had purchased from plains ranchers who had rounded up the bison and raised them as novelties. Jones designed his little corral of bison in Yellowstone to attract tourists. He located the corral near Mammoth Hot Springs—not a site preferred by the few wild bison remaining in the park, but near to the park's busiest entrance and therefore most accessible to visitors. Near the corral, Jones established a private museum and souvenir shop. The park superintendent accused Jones of "selling mementoes to tourists and generally prostituting himself for commercial gain." Jones' herd resembled another established in the park in 1896 by E.C. Waters, who operated a steamboat that ferried tourists across Lake Yellowstone. To lure passengers, Waters purchased four bison from a Texas rancher, installed them on an island in Lake Yellowstone, and, until 1907, operated a "game show" there (Isenberg 1997: 186).

For Eastern United States preservationists such as the members of the American Bison Society, however, Jones' style of wildlife protection was ideal: it assured tourists in Yellowstone a glimpse of the bison. The American Bison Society imagined bison preserves as places for urbanites to visit in order to experience the restorative effects of wilderness. Founded in New York City in 1905, the society was both geographically and economically exclusive. In 1908, 79 percent of its members lived in New York, New Jersey, Pennsylvania, and New England. Like the Poppers, the members of the American Bison Society were educated, elite, urban, and far away from the Great Plains and the bison (181-82).

The members of the American Bison Society did not romanticize the natives who had once subsisted on the bison. Rather, they lionized hunters-turned-game-wardens such as Jones or cowboys-turned-bison-herders such as Frank Rush, who enthusiastically welcomed the establishment of a federal bison reserve in western Oklahoma. Rush said in 1907: "The cowboy is rapidly becoming as extinct as the buffalo.... Back there on the range these buffalo will be attended to by some of the old cowboys who hunted buffalo on the plains in pioneer days" (Isenberg 1997: 182). Thus, the preservation of one icon of the old West—the bison—would help to preserve another—the cowboy—at the very time when prominent Americans feared that the disappearance of frontier conditions threatened American culture.

The members of the American Bison Society fixated on figures such as Jones and Rush because they agonized over what they perceived to be the waning of American masculinity. William Temple Hornaday, a zoologist who led the American Bison Society, called the bison "as dangerous as a lion" and praised Jones for his "heroism" in wrangling them (Isenberg 1997: 172). Indeed, the preservation of the bison was a decidedly gendered concern. In 1908, over 85 percent of the members of the American Bison Society were men. This preponderance of men was by no means typical of preservationist or conservationist organizations. Many women were enthusiastic participants in other such groups (Merchant 153-70). Women had comprised much of the membership of the animal anti-cruelty societies that in the 1870s had lobbied—unsuccessfully—against the wasteful slaughter of bison. The failed anti-cruelty campaign was an effort to extend the nineteenth-century feminine ethic of compassion to the treatment of animals by hunters, drovers, and teamsters (Isenberg 2000: 144). By contrast, the successful bison preservation campaign thirty years later depicted the salvation of the bison as a task that demanded manliness. Only by embracing the masculine ideology of frontier conquest did Euroamericans invest enough cultural significance in the bison to warrant its preservation.

Preservationism unintentionally confirmed the nineteenth-century transformation of the bison from a species with continuous habitation of the plains to a fragmented population occupying disjunct habitats. The fragmentation of the herds had a deleterious effect on the genetic diversity of the bison. All bison in North America are descended from the roughly five hundred survivors of the commercial slaughter of the nineteenth century—a so-called "bottleneck" in the transfer of genes. The exile of bison to small, dispersed preserves exacerbated this homogeneity (Berger and Cunningham 10). In the first decades of the twentieth century, preservationists could not have known all of this. Yet they regarded the domestication of the bison not as regrettable but as laudable. Convinced that the bison's future as a tourist attraction was assured, the society was unmoved in 1920 when in Utah the owner of a herd of over two hundred bison disposed of his animals by selling the right to hunt them for $250 a head. Hornaday wrote that "inasmuch as the owners ... find them an unbearable nuisance and an interference with their cattle-growing operations, what else is to be done than to get rid of them?" In 1922, the society acquiesced to the slaughter of surplus bison in Yellowstone and at Buffalo National Park in Canada. Another member of the society wrote that "it is a matter of gratification that the buffalo are becoming so numerous in some of the government herds that it presently will become necessary to treat the surplus bulls as so many domestic cattle" (Isenberg 1997: 189-90).

By the 1920s, the saviors of the bison echoed its slaughterers of the nineteenth century in calling for the killing of bison in order to open the North American West to livestock. They called for a sustainable harvest of the bison, of course—an obvious and crucial distinction. That difference, however, should not obscure the continuities between the destruction of the bison in the nineteenth century and its preservation in the twentieth. In both eras, Euroamericans domesticated the bison and its environment, transforming them to suit the convenience of an industrial economy.

One hundred years ago, the frontier historian Frederick Jackson Turner understood the domestication of the North American environment as a transformation from wilderness to civilization. Aware of the ecological costs of modernization in ways that Turner was not, we now regard it with considerably less triumphalism than he did, recognizing that we are all embedded in nature, and that the demise of any species ultimately affects everything else (including us) in nature's interconnected system. The easy response to this recognition would be to embrace the anti-modernism of romantic noble savagism or masculine frontier preservationism. But if people are embedded in nature, they are also embedded in their historical context. We cannot become like those people of the past; we can only try (imperfectly) to understand them. Part of that understanding is the recognition that both the destroyers and the saviors of the bison were, in strangely similar ways, implicated in the domestication of the Great Plains.

The past is also not a menu from which we can order *à la carte* historical environments. Reconstructing a bygone environment is an inherently speculative enterprise in which the best we can manage are educated guesses (Hall 240). Questions that are fundamental to a Buffalo Commons (how many bison inhabited the historic Great Plains?) lack definitive answers. If one restores bison to the Great Plains, ought one not also restore wolves, the bison's primary predator? Moreover, any restoration project requires that one select a point in the past that represents an ideal toward which the restoration project will strive. Should a Buffalo Commons aim to restore the Great Plains to its condition in 1803, at the time of the Louisiana Purchase, when tens of thousands of natives mounted on horses hunted the bison? Or 1492, before Europeans brought horses to the continent? Or 13,000 BC, before the arrival of prehistoric immigrants from northwestern Asia, when the Great Plains was populated by saber-tooth cats, mammoths, and giant bison (a much larger relative of the modern "dwarf" bison)? The last idea, called "Pleistocene rewilding," was proposed by the ecologist Josh Donlan in 2005. Donlan began by making the same observation that the Poppers made in 1987: the Great Plains' human population is declining rapidly, offering an opportunity for environmental renewal. While the Poppers viewed the Dust Bowl as the signature catastrophe of the region, Donlan looked farther back into prehistory; he sought to redress the extinction of Ice Age megafauna. Rather than re-introduce bison to the Great Plains, however, Donlan called for introducing modern proxies for the extinct mammoths and saber-tooth cats: elephants and lions imported from Africa (Donlan 913-14).

Bison were never as stable or enduring a presence in the grasslands as we once thought, but in the context of the environmental history of this region since the 1880s, the bison seem stable and enduring by comparison. The Great Plains has endured three environmental disasters in little over half a century: the near extinction of the bison, the collapse of the free-range cattle industry in the 1880s, and the Dust Bowl of the 1930s. Another potential catastrophe—the exhaustion of underground aquifers used to irrigate crops—looms in the future. Returning bison to a large part of the plains will not return us to Eden. But over the last few thousand years, we have learned that a grassland inhabited by bison, rather than grazed by cattle or planted with grain or soybeans, has proved to be the most sustainable regime.

That a sizeable bison preserve in the Great Plains modeled on the national parks would be an extraordinarily popular destination for tourists seems a safe predication. The Hayden Valley in Yellowstone National Park, home to most of Yellowstone's bison and the United States' *de facto* Buffalo Commons, is clogged with tourists every summer. The percentage of Americans who visit national parks has fallen in the last two decades since the Poppers first proposed the Buffalo Commons idea, but increasing numbers of visitors from Europe and elsewhere have made up the difference.

Yet is eco-tourism the best we can do for the bison? Is the best use of the Great Plains a park to serve our needs for recreation and moral comfort? The history of the American Bison Society demonstrates, if nothing else, how limited that vision can be. The historic natives of the Great Plains did not live in perfect harmony with the bison, but they did a better job of creating a green culture than the twentieth-century stewards of the Great Plains. By improvising a response to the European ecological invasion that brought horses to North America, they reinvented themselves as bison-hunting nomads in the eighteenth century. They sustained themselves in this experiment for a century—not because they had a mystical harmony with nature, but because their resource strategy was well suited to the solar economy of the grasslands.

Some Great Plains natives are following the example of their ancestors and once more adapting their resource use to the solar economy of the grasslands—but to the challenges of the present, just as the eighteenth-century nomads adapted to the challenges of their time. The Lakotas of the Rosebud, Pine Ridge, and Flandreau reservations, for instance, are all currently developing wind power projects. The United States already has a wind power capacity of 35,000 megawatts, which makes it the leading wind-power producing country in the world (Germany is second). Yet wind power accounts for only 2 percent of electricity generated in the United States. The Great Plains are the most economically feasible region to construct a more extensive wind energy regime in the United States. The Department of Energy's National Renewable Energy Laboratory estimates that by developing the wind power of the Great Plains and adjoining states, the United States could generate enough electricity so that by the year 2024, 20 percent of American electricity would come from renewable sources—a first step toward ending an unsustainable reliance on fossil fuels (Behr A26).

Like the natives of the eighteenth century, who became nomads in response to the crisis of the European ecological invasion, these natives of the twenty-first century are improvising a response to our current energy crisis by adapting to the solar economy of the Great Plains. Wind power, solar power, and biofuels—energy strategies that tap the power of the sun—all have a place in the solar economy of the North American grasslands. In the fossil fuel age, underground mineral resources have been critical. One of the initial limitations to the exploitation of solar energy is space. With current technologies, wind turbines, photovoltaic cells, and the cultivation of plants for biofuels require considerable land area. Space, however, is one thing the Great Plains has: it is 1,300,000 square kilometers in size—larger than Italy, France, and Germany combined. The Great Plains' potential for wind energy is so enormous that wind power consortiums in the eastern United States fear that cheaply produced Great Plains wind power will drive them from the market (Behr A26).

Some of those same native groups that are developing commercial wind power facilities, including the Pine Ridge Lakota, are also restoring bison to parts of their reservations. Indeed, currently, more bison can be found on tribal reservations than in federal parks and refuges. The tribal reintroduction projects are indisputably attempts to preserve cultural heritage, not unlike the efforts of the outdoorsmen and hunters of the American Bison Society a century ago. Restoration efforts are invariably entangled in nostalgia, cultural identity, and romantic tradition. Yet at the same time, the shortgrass-bison biome that these tribal groups, the Poppers, and others seek to restore was the most stable arrangement in the Great Plains' troubled environmental history.

The Pine Ridge Lakota's dual efforts at bison reintroduction and commercial wind power point the way toward a different sort of Buffalo Commons than the ones Catlin, the American Bison Society, and the Poppers imagined: bison grazing an extensive, unbroken grassland while wind turbines spin overhead. Such a regime would not mean that Great Plains inhabitants would be living in harmony with nature any more than the natives of the eighteenth and nineteenth century lived in perfect harmony with the grasslands—environments and human societies are both too changeable and unpredictable for that. Advocating such a use of the land does not represent a clear-eyed view of the best use of the land any more than the preservationists of the last century understood best how to manage the environment. Their notion (or anyone's notion, for that matter) of the best use of the land reflected their cultural precepts. Conservationists of the last century confidently manipulated environments—damming rivers, exterminating "varmints" such as coyotes and prairie dogs, suppressing forest fires—with humbling results. No one ought to embark upon a bison-wind power program in the Great Plains with similar hubris. Our understanding of the environment will never be perfect, and both that understanding and the environment itself will always change. That environments are changeable and cultures conflicted should not mean, however, that our response to the problems of the Great Plains—problems the Poppers correctly identified in 1987—should be cynicism or resignation. It means, instead, that solutions to those problems will be neither facile nor final.

Works Cited

Behr, Peter. "Plentiful Great Plains Power Blows in Opponents from All Corners." *New York Times* 8 March 2010.

Berger, Joel and Carol Cunningham. *Bison: Mating and Conservation in Small Populations.* New York: Columbia University Press, 1994.

Callenbach, Ernest. *Bring Back the Buffalo! A Sustainable Future for America's Great Plains.* Washington, DC: Island Press, 1996.

Catlin, George. *North American Indians: Being Letters and Notes on their Manners, Customs, and Conditions, Written During Eight Years' Travel Amongst the Wildest Tribes of Indians in North America, 1832-1839.* Vol. 1. Philadelphia: Leary, Stuart & Co, 1913.

Caughley, Graeme. "Wildlife Management and the Dynamics of Ungulate Population." *Applied Biology.* Ed. T.H. Coaker. London: Academic Press, 1976. 180-240.

Crosby, Alfred W. *Ecological Imperialism: The Biological Expansion of Europe, 900-1900.* New York: Cambridge University Press, 1986.

Dillehay, Tom. "Late Quaternary Bison Population Changes in the Southern Plains." *Plains Anthropologist* 10 (1974): 180-96.

Donlan, Josh et al. "Re-wilding North America." *Nature* 436 (18 August 2005): 913-14.

Dowd, Mark. *Conservation Refugees: The Hundred-Year Conflict between Global Conservation and Native People.* Cambridge, MA: MIT Press, 2009.

Flores, Dan. "Bison Ecology and Bison Diplomacy: The Southern Plains from 1800 to 1850." *Journal of American History* 78 (1991): 465-85.

Forsythe, David M. and Peter Caley. "Testing the Irruptive Paradigm of Large-Herbivore Dynamics." *Ecology* 87 (2006): 297-303.

Gabler, Robert E., Robert J. Sager, Sheila M. Brazer, and Daniel L. Wise, *Essentials of Physical Geography.* 3rd ed. Philadelphia: Saunders, 1987.

Gates, Paul W. "Land Policy and Tenancy in the Prairie States." *Journal of Economic History* 1 (1941): 60-82.

Goudie, Andrew S. *The Nature of the Environment.* 4th ed. Oxford: Wiley-Blackwell, 2001.

Hall, Marcus. *Earth Repair: A Transatlantic History of Environmental Restoration.* Charlottesville: University of Virginia Press, 2005.

Hornaday, William Temple. "Extermination of the American Bison, with a Sketch of its Discovery and Life History." *Annual Report of the Smithsonian Institution, 1887.* Vol. II. Washington, DC: Government Printing Office, 1889.

Hughes, J. Donald. *American Indian Ecology.* El Paso: Texas Western University Press, 1983.

Ingold, Tim. *Hunters, Pastoralists, and Ranchers: Reindeer Economies and their Transformations*. Cambridge: Cambridge University Press, 1980.

Isenberg, Andrew C. *The Destruction of the Bison: An Environmental History, 1750-1920*. New York: Cambridge University Press, 2000.

———. "The Returns of the Bison: Nostalgia, Profit, and Preservation." *Environmental History* 2 (1997): 179-96.

Kelly, Fanny. *My Captivity Among the Sioux*. New York: Citadel, 1933.

Krupnik, Igor. *Arctic Adaptations: Native Whalers and Reindeer Herders of Northern Eurasia*. Hanover, NH: University Press of New England, 1993.

Lee, Richard B. "What Hunters Do for a Living, or, How to Make Out on Scarce Resources." *Man the Hunter*. Ed. Lee and Irven DeVore. Chicago: University of Chicago Press, 1968. 30-43.

Matthews, Anne. *Where the Buffalo Roam: Restoring America's Great Plains*. Chicago: University of Chicago Press, 2002.

McHugh, Tom. *The Time of the Buffalo*. Lincoln: University of Nebraska Press, 1972.

Melville, Elinor. *A Plague of Sheep: Biological Consequences of the Conquest of Mexico*. New York: Cambridge University Press, 1994.

Merchant, Carolyn. "The Women of the Progressive Conservation Crusade, 1900-1915." *Environmental History: Critical Issues in Comparative Perspective*. Ed. Kendall E. Bailes. Lanham, MD: University Press of America, 1985. 153-70.

Petrie, Matthew D. "Climate Forcings and the Nonlinear Dynamics of Grassland Ecosystems." Master's thesis. University of Kansas, 2010.

Pfister, Christian. "The Early Loss of Ecological Stability in an Agrarian System." *The Silent Countdown: Essays in European Environmental History*. Ed. Christian Pfister and Peter Brimblecomb. Berlin: Springer-Verlag, 1990. 37-55.

Popper, Deborah Epstein and Frank J. Popper. "The Buffalo Commons: Metaphor as Method." *Geographical Review* 89 (1999): 491-510.

———. "The Great Plains: From Dust to Dust." *Planning* 53 (1987): 12-18.

Roe, Frank Gilbert. *The North American Buffalo*. Toronto: University of Toronto Press, 1951.

Seton, Ernest Thompson. *Lives of Game Animals*. Vol. 3. New York: Doubleday, 1929.

Shannon, Fred A. "A Post-Mortem on the Labor-Safety-Valve Theory." *Agricultural History* 19 (1945): 31-37.

Turner, Frederick Jackson. "The Significance of the Frontier in American History." *American Historical Association Annual Report* (1893): 199-227.

Turner, James M. "The Promise of Wilderness: A History of the American Environmental Movement, 1964-1994." PhD diss. Princeton University, 2004.

Worster, Donald. *Dust Bowl: The Southern Plains in the 1930s*. New York: Oxford University Press, 1979.

Zarrilli, Adrián Gustavo. "Capitalism, Ecology, and Agrarian Expansion in the Pampaean Region, 1890-1950." *Environmental History* 6 (2001): 561-83.

Facing *The Day After Tomorrow*: Filmed Disaster, Emotional Engagement, and Climate Risk Perception

Alexa Weik von Mossner

In his 2009 address to the IPCC, UN Secretary-General Ban Ki-moon expressed his concerns about the risks associated with climate change in perhaps somewhat unexpected terms. The scenarios outlined in the 2007 IPCC report, Ban declared, "are as frightening as a science fiction movie, but they are even more terrifying, because they are real" (Ban). Ban thus introduced his call for a new environmental ethics with an allusion to popular culture, offering a seemingly concrete referent for an abstract scientific scenario. Ban's resorting to science fiction in his attempt to communicate the urgency of the current environmental crisis points to the difficulties that people experience when trying to imagine the potentially catastrophic outcomes of their current lifestyles. Understanding the implications of local and global environmental risk requires not only knowledge and awareness, but also imagination. The wide distribution of scientific studies and assessments can certainly help raise awareness among the general public, and has done so over the past decades. However, as social science scholars David Lewis, Dennis Rogers, and Michael Woolcock have recently argued with respect to "The Fiction of Development" (2008), imaginary narratives can communicate knowledge about social or economic issues in ways that are different but often just as valuable as scientific or scholarly studies: "Not only are certain works of fiction 'better' than academic or policy research in representing central issues relating to development, but they also frequently reach a wider audience and are therefore more influential" (198). The same is true, I believe, for "the fiction of climate change," and another important strength of such imaginary narratives is that they are much better at engaging emotions, especially (but not exclusively) when the imaginary narrative in question is popular film.

The "science fiction movie" Ban Ki-Moon had in mind when making the aforementioned statement is in all likelihood Roland Emmerich's 2004 blockbuster *The Day After Tomorrow*, a film that aptly combines features of the melodrama and the disaster narrative to engage its viewers cognitively and emotionally in a spectacular story about abrupt climate change. Produced by Emmerich's own company Centropolis and the Canadian studio Lionsgate with a budget of $125 million, and distributed by Twentieth Century Fox, it achieved a total gross revenue of $652 million. According to Box Office Mojo this makes Emmerich's film the second-highest grossing film of all times in the category "Environmentalist," behind James Cameron's *Avatar*. Emmerich's film also has made it to rank 3 in the category "Controversy," however, beaten only by *The Passion of Christ* and *The Da Vinci Code*. This latter achievement reflects the wide attention the film received not only from reviewers and journalists, but also from climatologists, sociologists, environmentalists, and American government officials, who either lauded, criticized, or vilified the film. While certainly not the first cultural text with a significant effect on the general public in the United States and beyond, it was the first popular film to be credited with—and chastised for—turning public awareness to the issue of climate change .[1]

This social and political impact, I will argue in the following, is to a large degree due to the fact that *The Day After Tomorrow*—like all disaster films—appeals to both rational thinking and emotions as it tells its tale of abrupt and catastrophic climate change. Emmerich transforms abstract scientific scenarios into a concrete story about a specific place and particular people, and he turns current perceptions of risk—anticipated catastrophes, as Ulrich Beck calls them—into audio-visual spectacles that have a direct *visceral* effect on the viewer (see Beck 9). Such a fictional concretization of abstract notions into emotionally engaging stories is important, because as Paul Slovic and other psychologists working in the area of risk perception have found out, emotions matter at least as much as analytical thinking in both risk perception and decision making. Building on the work of Antonio Damasio and other scholars working in the field of neuroscience and behavioral neurology, Slovic explains in *The Perception of Risk* that over time, he and his colleagues "have come to recognize just how highly dependent [risk perception] is upon intuitive and experimental thinking, guided by emotional and affective processes" (xxxi). Although deliberation and analysis are important factors in many decision-making circumstances, says Slovic, "reliance on affect and emotion is a quicker, easier, and more efficient way to navigate in a complex, uncertain, and sometimes dangerous world" (xxxi). In a 2004 article entitled "Risk as Analysis and Risk as Feeling," Slovic et al. go even further, stating that "analytic reasoning cannot be effective unless it is guided by emotion and affect" (313) and that "we cannot assume that an intelligent person can understand the meaning of and properly act upon even the simplest of numbers such as amounts of money or numbers of lives at risk … unless these numbers are infused with affect" (321). This builds directly on Damasio's claim, in his groundbreaking *Descartes' Error*, that "emotion and feeling, along with the covert physiological machinery underlying them, assists us with the daunting task of predicting an uncertain future and planning our

[1]   Davis Guggenheim's *An Inconvenient Truth* was only released in May 2006.

actions accordingly" (xxiii). If affect is so central to both the perception of risk and to decision making in the face of such an enormous risk as climate change, an entertainment form that not only reaches millions of people the world over but also succeeds in engaging them emotionally certainly deserves closer attention.

Given the exciting new work done in recent years by cognitive film scholars on emotions and film structure, we are now able to investigate more theoretically the notion that films have the ability to engage emotions. Drawing on the work of Noël Carroll, Ed Tan, Carl Plantinga, and others, who, like Slovic and his colleagues, are interested in the relationship between perception, emotion, and cognition, I will investigate in the following how, exactly, a blockbuster film like *The Day After Tomorrow* engages its (cross-cultural) audiences emotionally while offering them along the way a few lessons about some of the potential dangers of abrupt climate change. That audiences across the world *have learned* some of these lessons as a result of watching this particular disaster movie has been shown in five independently conducted studies in the United States, Britain, Germany, and Japan. The important qualitative research that has been done in these studies is what makes *The Day After Tomorrow* a particularly interesting case. As Fritz Reusswig, the lead author of the German study, points out, the really interesting fact about those studies is that they can offer us some empirical data on the question of "if and how a global media event like the simultaneous launch of TDAT in almost 80 countries across the globe … has affected the public with different cultural and political backgrounds in different countries" (2004b: 1).

The social and political relevance of cultural texts is no news to literary and cultural studies scholars; however, they rarely have the privilege of having their findings substantiated by empirical research done by scholars in institutions like the Tyndall Centre for Climate Change Research, the Yale School of Forestry & Environmental Studies, and the Potsdam Institute for Climate Impact Research. This chapter aims at drawing connections between these audience response studies on the one hand, and cognitive approaches to film studies on the other, in order to develop a more comprehensive understanding of how, exactly, *The Day After Tomorrow* interacts with its viewers' emotions, perceptions, and cognitions. Emmerich's film, I will show, engages its viewers emotionally in both its melodramatic plotline and the mind-boggling spectacle of "natural" disaster. The basic information about climate change it communicates between the lines is thus infused with affect and likely to make an impact on viewers' perception and cognitive understanding of the issue.

Responding to *The Day After Tomorrow*: The Reception Studies

Cultural texts do not exist in a vacuum, and when filmmakers imagine environmental risk, they do so in culturally and historically specific ways. Such contextual factors, as reception theorist Janet Staiger reminds us, also account to a large degree "for the experiences that spectators have watching films … and for the uses to which those experiences are put in navigating our everyday lives" (1). As Geoff King points out, "between viewer and text come numerous other mediations and meaning-creating factors," including commercial imperatives, and broader social and historical contexts (7). Cultural, social, historical, and economic factors profoundly circumscribe what filmmakers can imagine and what audiences can interpret and learn, and "so do more particular contextual factors such as class, gender, or racial background, or narrower group or personal histories" (King 7). This is one of the reasons why audience response studies are particularly interesting in cross-cultural comparison. It is thus helpful that the five empirical studies on *The Day After Tomorrow* were conducted more or less simultaneously in four different countries, examining the attitudes of audiences toward global warming before and after seeing the film, and, in one case, also comparing them to the attitudes of non-viewers.[2]

The American study was conducted by Anthony Leiserowitz—a risk perception scholar and the current director of the Yale Project on Climate Change—and published in the November 2004 issue of *Environment*. Leiserowitz is interested in the effects of the massive press controversy that predated and accompanied the release of *The Day After Tomorrow* on the one hand, and in the attitudes of viewers before and after seeing the film on the other. With regard to the first concern, Leiserowitz notes that

> some commentators feared that the catastrophic plotline of *The Day After Tomorrow* would be so extreme that the public would subsequently dismiss the

---

[2]   Fritz Reusswig notes in his article on "The International Impact of *The Day After Tomorrow*" that the five independently conducted studies differ in methodology and approach, but asserts that a comparative view can nevertheless offer some relevant conclusions. In another paper, entitled "Climate Change Goes Public," Reusswig gives the following information about the five studies: all but the American study were based on questionnaires. The American study made use of a web-based survey. Two studies (Reusswig et al. and Lowe et al.) also used focus groups. The size of the sampling was N=1118 (only filmgoers) for Reusswig et al's study in Germany, N=384 (only filmgoers) for Aoyagi-Usui's study in Japan, N=301 for Lowe et al's study in the UK, N=200 (only filmgoers) for Balmford et al's study in the UK, and N=529 (filmgoers and general public) for Leiserowitz's study in the United States. For additional information see Fritz Reusswig "Climate Change Goes Public."

entire issue of global warming as fantasy … others spun a scenario in which, panicked by the movie, the U.S. public would force Congress to pass climate change legislation, President George W. Bush would subsequently veto the bill, and challenger John Kerry would exploit public hysteria over global warming to win the U.S. presidential election. (23)

There is reason to believe, however, that the most "hysterical" reactions were actually those of the Bush administration itself. Leiserowitz mentions a leaked memo from NASA administrators (which was later published by the *New York Times*), which stated that "no one from NASA is to do interviews or otherwise comment on anything having to do with" the film, and that "any news media wanting to discuss science fiction vs. science fact about climate change will need to seek comment from individuals or organizations not associated with NASA" (Revkin). The senior NASA scientist who provided the *New York Times* with this confidential memo reportedly said that he had done so because "he resented attempts to muzzle climate researchers" (Revkin). Regardless of this attempted muzzling of federal climate scientists (which was later partially retracted), the impending release of Emmerich's film received ample attention in the US media. Leiserowitz's study reveals that the film generated "more than 10 times the news coverage of the 2001 IPCC report," and, given that "a key component of the risk amplification process is media attention" this is an important factor in the film's social impact (34). However, even as that sounds considerable, Leiserowitz reminds us, "relative to other news stories, global warming is [still] a rarely reported issue" (34). At the time when he collected his data, "the Abu Ghraib prison scandal had in turn more than 10 times the coverage of *The Day After Tomorrow*" (34) and thus 100 times the news coverage of the IPCC Report.

With regard to viewer's attitudes toward climate change before and after seeing the movie, Leiserowitz also comes to clear and perhaps somewhat surprising conclusions. All parameters considered, he summarizes the results of his study thus;

> *The Day After Tomorrow* had a significant impact on the climate change risk perceptions, conceptual models, behavioral intentions, policy priorities, and even voting intentions of moviegoers. The film led moviegoers to have higher levels of concern and worry about global warming, to estimate various impacts on the United States as more likely, and to shift their conceptual understanding of the climate system toward a threshold model. Further, the movie encouraged watchers to engage in personal, political, and social action to address climate change and to elevate global warming as a national priority. (34)

As far as audience response goes this is pretty impressive. And it is not a purely American phenomenon either. The international impact of *The Day After Tomor-*

row was, as Fritz Reusswig from the Potsdam Institute for Climate Impact Research notes, somewhat but not entirely different. The changes in attitudes of the German audiences that Reusswig surveyed in his own study were less dramatic, which he ascribes to a large degree to the fact that in this case a large portion of the audience was already sensitized to the issue of climate change, so much so, in fact, that it was one of their main reasons for watching the film (this was different for the American audience, which was primarily interested in seeing Emmerich's latest disaster movie). Nevertheless, Reusswig concludes that the film *did* have significant effects on its German viewers and that "the entertainment industry seems to have done quite a lot for the public awareness of climate change" (2004a: 43).

The two British studies came to similar conclusions. "Overall," Andrew Balmford et al. note, "our findings confirm that intense dramatizations have real potential to shift public opinion" (1713). However, the researchers add, "the question remains whether such portrayals can be made more accurate (and thereby less confusing) without losing their popular appeal" (1713). Lowe at al, the authors of the second British study, conclude that "our research shows that seeing the film, at least in the short term, changed people's attitudes; viewers were significantly more concerned not only about climate change, but also about other environmental risks such as biodiversity loss and radioactive waste disposal" (2). The Japanese study, which was conducted by Aoyagi-Usui at the National Institute for Environmental Studies (NIES), has remained unpublished, but Reusswig reports in his comparative overview of the five studies that Japanese interviewees actually perceived abrupt climate change scenarios as *less* likely after seeing the film, while showing at the same time a higher motivation to take individual action to prevent it (2004b: 2). Reusswig concludes that "the success and impact of 'The Day After Tomorrow,' as explored by the five studies, indicates that we should be more attentive to the issue of awareness raising and education—and slightly more optimistic too" (2004b: 3). Thomas Lowe's final judgment about the film, published one year later, is less positive. Despite the film's potential "to tap into the accessible parts of psychological function," he writes, it "falls short of being the ideal risk communication tool by departing from the realms of reality and failing to offer audiences a basic understanding of causes and measures for mitigation" (2006b: 5).

It is probably safe to say that Emmerich never planned to create "an ideal risk communication tool," but Lowe nevertheless has a point.[3] *The Day After Tomor-*

---

[3]   Lowe's formulation is in fact somewhat problematic here, since it seems to suggest that there might be such a thing as an "ideal risk communication tool" to begin with. However, as Ulrich Beck points out in *World at Risk*, risk perception is a highly complex issue, which is why it is a problem that risk communication is often imagined as an information flow from rational,

row significantly departs from the realms of reality and science, and it depicts a scenario that is far beyond mitigation. The question remains, then, why the film has nonetheless been moderately successful in raising awareness for climate change risks among its audiences, and why, as Leiserowitz asserts, it even encourages watchers to engage in personal, political, and social action. Paradoxically, a good part of the answer to that question is that *The Day After Tomorrow* comes across as pure entertainment, with no apparent pedagogical agenda to push, but with a strong *visceral* and *emotional* impact on its audience. Before I turn to the film itself I want to outline some of the very useful and convincing ways in which film scholars who use cognitive approaches have theorized the emotional structure of film and its impact on the viewer.[4]

The Affect of Cinematic Narration: Cognitive Approaches to Film Emotions

Movie theaters, as film scholars Carl Plantinga and Greg M. Smith remind us, occupy a central place "in the emotional landscape of the modern world as one of the predominant spaces where societies gather to express and experience feelings" (1). Not only is much of our experience of films saturated with emotion, but, as Noël Carroll notes, "our emotional engagement constitutes, in many instances, the most intense, vivid, and sought-after qualities available in the film experience" (24). Movies, Carroll argues, are "objects that are well constructed to elicit a real emotional response from our already existing emotion systems" (23) and the emotions we experience when watching a tear-jerking melodrama, a scary horror film, or an arresting suspense drama are thus not really different in kind from the 'real' emotions we have in our everyday lives. What is different is that in life there is much less guidance or direction for our emotions. Our perception, as well as our emotion

scientific experts to irrational, subjective laypeople. Such an understanding, Beck explains, tends to bracket and dismiss important factors, "such as different forms of non-knowing, contradictions among different experts and disciplines, ultimately the impossibility of making the unforeseeable foreseeable" (12). Lowe seems to have noticed the problem inherent in his overly strong claim. In his "Is This Climate Porn?," published several months later, he uses an almost identical formulation but speaks of a "conventional risk communication tool" (5) rather than an "ideal risk communication tool."

[4] Not all film scholars who assume a cognitive perspective are interested in the affective dimension of filmic narration. For good introductions to the field and the role of psychological concerns in it see David Bordwell, "Cognitive Theory" (2009); Carl Plantinga, "Cognitive Film Theory: An Insider's Appraisal" (2002); Carl Plantinga, "Affect, Cognition, and the Power of Movies" (1993).

system, is constantly confronted with a massive array of largely unstructured data, and when we react emotionally to a real-life situation, we have to filter out the relevant elements from this constant information overload. When we watch a film, Carroll explains, the situation is different. Here, "the filmmakers have already done much of the work of emotionally organizing scenes and sequences for us through the ways in which [they] have foregrounded what features of the events in the film are salient" (Carroll 28). Films *guide* our emotions while we watch them; as Greg Smith puts it, they continually "offer [us] invitations to feel" in certain ways (12). We can accept those invitations and "experience some of the range of feelings proffered by the text," or we can reject them (G. Smith 12). If we choose to accept them, the film becomes a full body experience, because emotions, as Carroll explains, involve both cognition and "feelings," the latter being "sensations of bodily changes, like muscle contractions," or the welling of tears (Carroll 24).

Drawing on the insights of cognitive psychology and neuroscience rather than on the methods of psychoanalytical research, cognitive film scholars like Carroll, Plantinga, and Smith argue that there is really no strict division between rational and emotional cognition, and that the spectator's faculties of cognition and judgment are in fact of central importance in the process of eliciting an emotional response to film.[5] As Plantinga points out, "one cannot seriously examine emotional effect without considering perception and narrative comprehension" (2009b: 239). This also explains what Plantinga calls the "conditional realism" in spectator response: the viewer experiences emotions that are very similar to what she would feel in a "real" situation, with the difference that she is at the same time aware of the fact that what she watches is a *fiction* and that this fiction is *mediated* (see 2009b: 239). As communication scholars Busselle and Bilandzic put it, "the experience of comprehending and engaging with a narrative is complex and multifaceted, involving losing awareness of some aspects of the actual world and gaining awareness of both cognitive and emotional aspects of an alternative world" (Busselle and Bilandzic). Also, as both Plantinga and Busselle and Bilandzic point out, the viewer will usually have an understanding of genre conventions and bring other previous knowledge to the film which might amplify her emotions or interfere and conflict with them.

Central to the emotional engagement of the audience in narrative films is, generally, the main character or hero of the story, but sympathy or antipathy for protagonists or villains are not the only emotions that keep the viewer engaged in the

---

[5]   David Bordwell and Noël Carroll's edited volume *Post-Theory: Reconstructing Film Studies* is very helpful for a better understanding of how cognitive approaches challenge the long-standing dominance of poststructuralist and psychoanalytical approaches.

plot of a film. Plantinga identifies three other categories of emotions that films elicit in viewers: direct emotions, which are "responses to the narrative and its unfolding;" artifact emotions which are directed at the film as an artifact; and meta-emotions, which are aimed at the spectator's own responses or those of other spectators" (2009b: 242). These emotional responses are often mixed and tend to interact or interfere with each other, and they are necessarily different for different viewers in different audiences. Nevertheless, the film text itself offers its viewers certain emotional positions—invitations to feel, as Greg Smith would put it—and I want to dedicate the last part of this chapter to working out at least some of the countless "invitations" offered by Emmerich's *The Day After Tomorrow* and relating them to the film's emotional and cognitive effect on its audiences.

Melodrama, Science, and Spectacle: The Emotional Appeal of *The Day After Tomorrow*

Emmerich's film opens in the vast white expanse of Antarctica. The first minutes of a film set the tone for the rest of the story, and the first scenes of *The Day After Tomorrow* serve several purposes: firstly, a high-angle tracking shot—basically a simulated helicopter shot—introduces us to the breathtaking beauty of the polar region—a "natural" beauty, one should keep in mind, which is wholly computer-animated. When we finally see a group of (equally animated) humans, they look tiny and insignificant in this vast landscape, but as we get closer we realize that they are not, actually, lost. As the prominent American flag indicates, the isolated humans on the Larsen B Ice Shelf are a group of American scientists doing ice core drillings, and the second purpose of these first minutes of the film is the introduction of Jack Hall (Dennis Quaid), a paleoclimatologist who combines in one person the qualities of the melodramatic hero and the action man of the typical disaster film.[6] Without much ado, the film goes on to demonstrate Jack's action hero qualities. While his inexperienced colleague does the drilling, a gigantic fracture suddenly forms in the ice, leaving the scientists' little camp at the edge of an enormous abyss and their extracted ice cores on the other side. Without hesitation, Jack jumps across the abyss, salvages the ice cores and, with another, even riskier leap,

---

[6]   The screen-filling American flag not only eases the cut from animation to partial animation combined with life action, it also tells us clearly that this is going to be a story about the United States. The rest of the world really only exists as a cipher in Emmerich's film, despite his somewhat whimsical attempts to add a global dimension through the inclusion of a few scenes in Ireland, Japan, and India.

returns to his comrades. As the camera zooms out we witness the spectacular breaking off of an ice shelf "the size of Rhode Island" which foreshadows the coming disaster—the third purpose of this opening sequence.

The first and third of these purposes are fulfilled brilliantly: as spectators, we feel the emotional impact of both the beauty of nature (aided by the excellent musical score) and its destruction (aided by our previous knowledge about the effects of climate change).[7] The powerful evocation of a breathtakingly beautiful but suddenly also threatening and threatened natural landscape elicits awe in us for the sheer beauty of the images and sadness for a vulnerable ecological space. The second purpose of the opening scene, the introduction of the scientist as action hero, is arguably rather less successful. When Jack jumps heroically across the expanding crevasse in the ice cap to save the precious ice cores, most viewers are likely closer to laughter than awe. As paleoclimatologist William Hyde scathingly remarked in a blog after seeing the film, "the movie is at its most stunningly accurate in its portrayal of paleoclimatologists. Paleoclimatologists are notoriously brave and of course very fit. Nary a one of us would hesitate to jump a widening crevasse—twice—while wearing arctic gear—to recover some ice cores which would take 2-3 hours to re-drill" (Hyde). Even if we find the unrealistic nature of Jack's actions amusing rather than stunning, however, he is now established as a dedicated scientist who does not hesitate to go to physical extremes if he deems it necessary.

After his daring stunt in the melting ice of Antarctica, we next see Jack in the confined political space of a UN conference, explaining the scientific scenario of abrupt climate change to the assembly members and thereby supporting, as Sylvia Mayer has pointed out, "an anthropocentric environmental ethics that insists that human welfare depends on considerations of ecological processes" (106). This is the moment in which the film establishes a rational notion of risk, one that is very similar to the climate change risk scenarios we have heard about in the real world. However, Jack is confronted with ignorance and a strong belief in short-term profits, and he thus cannot succeed in warning the politicians of the world about the catastrophe that they are bringing about. With these first two sequences, his two vastly more powerful opponents are established: the American government—personified in the film by a vice president who bears a striking resemblance to Dick Cheney—and "nature," which will from now on become increasingly more hostile. The figure of Jack, who emerges as a powerless but morally righteous hero in the face of human ignorance and greed, asks us to empathize with his position,

---

[7]   For a cognitive perspective on the emotional effects of film music see Jeff Smith, "Movie Music as Moving Music: Emotion, Cognition, and the Film Score."

and this is easy for us because—as in any melodrama—we already sense that he is right and that his opponents are stupid, dangerous, and wrong.[8]

As we will soon realize, Jack is not only handsome, righteous, and "notoriously brave," he also has, as Matthew Nisbet has observed, a number of other qualities: "He drives a hybrid car ... shares a strong bond with his co-workers, and risks his life early on in the film to save his colleague.... The only downside to Quaid's character is that he is a workaholic. He is completely devoted to climate science while missing out on his family life" (Nisbet). Jack thus has a number of virtues and only one serious (and forgivable) flaw, which he will resolve to correct in the course of the story. These qualities are part of what makes him a typical melodramatic hero: a highly virtuous David-figure forced to struggle against an overpowering, Goliath-like enemy for the common good and the well-being of others. It does not take long until we realize, together with Jack, that the process he described to the UN Assembly will not in fact happen in 100 or 1000 years, but right now and very quickly. Thus the much more commanding of his two opponents—nature, which has considerable agency in the film—begins to determine the direction of the story, first with vicious tornadoes, and later with towering flood waves and enormous storm systems that allegedly suck down from the stratosphere air so cold that it freezes everything it touches.

These scenes of disaster are the "money shots" of the film and in all likelihood the main reason why Emmerich wanted to make it. They come with powerful images and sounds that have a strong visceral effect on the spectator, engaging her emotionally on the nonempathetic level.[9] Geoff King has suggested that one theme common to many disaster films "is that of 'natural' or elemental force breaking into the paved, built-up and 'civilized' ... or 'decadent' and 'artificial' worlds created by humans" and that "the principal targets for destruction are symbols of luxury, decadence [and] arrogance" (146). To witness the annihilation of famous cultural landmarks—the first twister in *The Day After Tomorrow* swiftly erases the Hollywood sign—seems to be immensely pleasurable for a large number of

---

[8] For detailed accounts of melodrama in film narratives, see Robert Lang, *American Film Melodrama* and John Mercer and Martin Shingler, *Melodrama: Genre, Style, and Sensibility*. For an interesting discussion of the function of melodrama in disaster films see Despina Kakoudaki, "Spectacles of History: Race Relations, Melodrama, and the Science Fiction/Disaster Film".

[9] I am taking this term from Tan and Frijda, who differentiate between empathetic "F emotions" and nonempathetic "A emotions." The first group consists of "responses in the fictional world" including "sympathy, compassion, and admiration." The second group of emotions includes the enjoyment of "the sight of a majestic landscape" and is independent from "the significance of the situation for the protagonist" (52).

moviegoers and one of the main reasons why they go to see a particular film. King hypothesizes that this pleasure may be related to the specific emotional effect of such scenes. In contrast to the order and coherence provided by narrative, he suggests, "moments of spectacle may offer ... the illusion of a more direct emotional and experiential impact" (36). However, King does not explain why spectators would only experience the "illusion" of such a "direct emotional impact." In the eyes of cognitive film scholars at least, the emotional impacts of cinematic spectacle are hardly an illusion. Rather, such spectacular scenes work directly on the spectator's emotion system, triggering real emotional responses such as pleasure and awe.

Ed Tan and Nico Frijda suggest that the nonempathetic emotions provoked by awe-inspiring imagery must not necessarily be "immediately associated with the film story's action or with an understanding of the protagonist's feelings" (62). Such imagery, which may portray "an environment in which one feels tiny and insignificant" or other visually and aurally overwhelming scenarios, can, according to Tan and Frijda, trigger two kinds of response: "On the one hand, the stimulus may be attractive and call forth fascination.... On the other, it may have a repellent quality, eliciting a tendency to shiver and look for shelter" (62). The disaster spectacle of *The Day After Tomorrow* arguably triggers both responses at the same time, which is part of what makes it so attractive to a mass audience. When American moviegoers who watched Emmerich's film "were found to have significantly higher risk perceptions than 'nonwatchers'" (Leiserowitz 26), this may have been at least partially a result of the nonempathetic emotions they experienced while watching the most spectacular scenes of film, emotions that include fascination, awe, and a vague sense of fear.

One complaint leveled at *The Day After Tomorrow,* not only by reviewers and other commentators but also by the authors of the five reception studies, is that its spectacular disaster scenarios depart significantly from the predictions of climatologists, and that they generally violate notions of scientific plausibility. However, as David Ingram has pointed out, we miss the point of Emmerich's film if we simply dismiss it for its scientific inaccuracies. As a science fiction film, Ingram explains, Emmerich's movie uses "realist elements of climate science as a starting point for melodrama and fantasy, so that it can dwell on the spectacle of extreme weather ... and also invite the audience's emotional engagement with the human interest story that becomes the main focus of narrative" (55). The human interest story is of course directly related to the spectacle of disaster and gains additional appeal from it. As Geoff King has observed, "narrative and spectacle can work together in a variety of changing relationships" (2) and "spectacle can have an impact similar to that of driving linear narrative: it has the potential to reinforce, almost physiologically, whatever the narrative asserts" (34).

This clearly is the case in *The Day After Tomorrow*. The narrative asserts that a failure to act in the face of tremendous environmental risk will lead to unprecedented human catastrophe, and the film's spectacle most powerfully reinforces this assertion. Since the unleashed natural forces are too powerful an opponent to be stopped or beaten, Jack's goals must focus on something that he actually *can* try to accomplish, albeit against tremendous odds: the timely evacuation of a part of the American population, and, to make the story personal and emotionally moving, the highly melodramatic rescue of his son Sam (Jake Gyllenhaal). Both of these goals are related, since Sam is in New York City and thus above the line that Jack draws on a map of the United States, indicating to the American government the divide between those that can be saved (the Southern half) and those that must be sacrificed (the Northern half).

In this highly emotional scene we again see a confrontation between the heroic scientist and the American government, but now the tables are turned. If Jack was the isolated outsider in the first scene (before the UN Assembly), now it is the ignorant vice president. The conflict-ridden expert-layperson relationship that emerges in this and in the earlier scene remains central to both of the film's main storylines. On the one hand, *The Day After Tomorrow* shows us a battle between Jack and other scientists (the experts) and the politicians (the laypeople) who have to make far-reaching decisions; on the other hand, we have a group of people in the New York Public Library who are confronted with conflicting information, most of it hearsay, and who also have to make existential decisions. In this group it is Jack's son Sam who takes on the role of the expert, since he has, through his father, privileged access to scientific knowledge. As we soon learn in the film, those who listen to the experts survive, while those who do not either drown or are later found frozen in the snow. The two plotlines and the moral messages they assert are thus directly—if sometimes crudely—linked to the film's spectacular disaster scenes, and the emotional appeal of the spectacle supports and reinforces the assertions of narrative.

Central to the emotional engagement of the audience in narrative films are, as I pointed out earlier, the protagonists of the story, and it is significant that in *The Day After Tomorrow* the main characters are all scientists or similar experts. Emmerich's scientists are not aloof nerds, but mildly flawed and highly emotional heroes, who seem to act irrationally because in the face of ecological risk they stick with their gut feelings and trust and belief in each other and in science. Jack is in love with his work and his main heroic act in the film—his long walk from Washington DC to New York in a subzero ice storm—is motivated solely by the promise that he made to his beloved son. Stephen Keane has argued that "disaster films are innately passive and survivalist (in the sense that when their central disasters occur the characters have no choice but to try to make their way up, down, or out to safety)" while "action movies are innately active and escapist" (53). Emmerich combines the two modes, however, and thus ends up with a main character who is motivated by love, and who is active rather than survivalist in the face of disaster. While Sam and his companions in the New York Public Library are passive in that they are forced by the disaster to fight for their survival, Jack's desperate attempt to save his son is an active decision, and his physical exposure to the storm a voluntary act that is based on moral obligation and love.

This highly emotional and melodramatic story is at the heart of the film, all filmic catastrophes and special effects notwithstanding. As Robin Murray and Joseph Heumann point out in their 2009 *Ecology and Popular Film*, the main plot of *The Day After Tomorrow* "revolves around Jack's quest to save his son and his son's evolution into a new eco-hero like his father ... Jack makes his heroic journey not to save the world—as we might expect an eco-hero and a climatologist to do—but to save his son" (10). The narrative invites the viewer to empathize with Jack and Sam, to see the fictional world through their eyes, and to feel sympathy for their goals and struggles. Carl Plantinga explains that sympathetic emotions typically "arise when the spectator assesses the narrative situation in response to a favored character's predicament and goals. When the viewer develops a concern that the goals of the character be met, this creates a desire for the attainment ... of the character's desired state or the escape from or avoidance of an aversive state" (*Moving* 88). *The Day After Tomorrow* asks us to care for Sam's goal of survival and of saving the girl he is in love with, and for Jack's goal to save as many Americans from the storm as he can and to keep his promise to his son. In face of a disaster of enormous proportions, Jack learns from his old mistakes and thus creates a new basis for his relationship with his son.

'Learning from one's mistakes' is a recurring theme in the film, both in the private and the political sphere, and it is clearly the message that the viewer is supposed to grasp. Catastrophes are often said to have transformative effects, and in *The Day After Tomorrow*, this certainly is the case. The greatest personal transformation and learning experience, however, is, surprisingly, not accomplished by Jack or his son, but by the man who so much looks like Dick Cheney, and who is now the new president of a United States that is as much changed as he himself is. In his first TV address to the nation after the disaster, he displays a new sense of humility and environmental ethics: "For years we operated under the belief that we could continue consuming our planet's natural resources without consequences. We were wrong. I was wrong. The fact that my first address to you comes from a consulate on foreign soil is a [testimony] to our changed reality." This is of course the central moment of catharsis and recognition, the moment that the film insists most blatantly on its environmentalist message. Even the new American president—previously Jack's opponent—has now learned from his mistakes and acknowledges that the mindless consumption of the earth's resources is unsustainable and morally wrong. It has only taken a few billion casualties for him to make that step.

Emmerich juxtaposes the presidential address with high angle tracking shots, which—as in the opening scene—show a white landscape (to the same musical score), with the difference that now that white landscape is New York City and the beautiful, peaceful white cover functions as a shroud that hides the millions of people lying dead underneath it. Like most makers of disaster movies, Emmerich makes sure that we do not dwell too much on the pain and death of the victims of the catastrophe and instead empathize with those who struggled and survived. Even as the film reaches its problematic happy ending, however, some things are shown to have drastically changed. As Sean Cubitt notes, *The Day After Tomorrow* builds "a sense of the unique as well as overwhelming character of disruption" (130). The monstrous nature at work in the middle of the film is stable and seemingly tamed again at the end, but America and the rest of the world are permanently trans-formed and order is *not* reestablished. Only utter devastation and disruption, Emmerich seems to suggest, can finally give humanity the ability to learn from its mistakes and change its unsustainable practices.

Conclusions

As its title suggests, Emmerich's film is set on the (ever shifting) "day after tomorrow"—that is, the near future. It is thus a piece of speculative fiction, or to put it in more familiar film terms, an eco-dystopian science fiction. As such, it stands in a longer tradition, one which, as Frederick Buell has pointed out, has historically "been anything but mere escapist fantasy. Just the opposite: it has not just reflected but influentially intervened in heated contemporary environmental-political dis-putes from its inception to the present day" (248). This is definitely true for Emmerich's film, regardless of its manifold flaws and shortcomings. As both Reusswig et al. and Lowe et al. point out in their studies, *The Day After Tomorrow* does not tell its audience anything about mitigation options; nevertheless, it "has trig-gered mitigation reactions in its public" (Reusswig et al. 58). In many ways, Reusswig et al. note, the film successfully relies on already existing perceptions of global warming risk—however vague they may be—and on its viewers' preexist-ing knowledge about mitigation measures.

While this is true, it is unlikely that the film would have been as successful in raising awareness had it only depicted a catastrophic scenario of abrupt climate change without engaging its audience emotionally in a hero's journey. As Thomas Lowe points out, "the moral sub-plot of the film, in which Jack Hall failed to be a proper father and wanted to make good his mistakes, could carry an important message. Perhaps it suggests to the viewer that society needs to think about its actions in general, not just with regard to the environment but to humanity as a whole" (2006b: 75). This is almost certainly part of the explanation as to why the film has been so effective in raising its viewers' climate risk perception. As viewers, Noel Carroll explains, we "tend to accept the projects of characters ... who strike us as virtuous" (40). And since "the efforts of the protagonists [tend to be] morally correct in accordance with the film's ethical system" (40), we tend to adopt that ethical system as ours, at least for the duration of the film. Jack Hall asks us to be concerned, with him, about the catastrophic outcomes of unmitigated climate change and for the lives of those people who will be directly affected by those outcomes. Although not discussing those risks critically or even correctly, *The Day After Tomorrow* thus succeeds in both of the main goals that Emmerich claims to have had when making the film: it has become a highly successful entertainment movie *and* it alerts and educates its audience about the real risks and dangers of climate change. As Leiserowitz notes, these divergent goals sometimes "coexist in an uneasy tension within the film" (26); however, these tensions seem not to interfere in the least with its success: at least in the United States, "70 percent of the [interviewed] moviegoers rated the movie as good or excellent" (26). Furthermore, Reusswig writes that "the impact studies of *The Day After Tomorrow* have entered a new, reflexive area of climate change research: the area of the impacts of impacts. Twentieth Century Fox Germany has established an initiative to facilitate emissions trading rights and reducing CO_2 emissions of services, events, and traffic" (2004a: 43). This reaction of a film studio to a scientific study can of course be dismissed as mere PR ploy, but given that the film industry emits significant amounts of CO_2 in the production and distribution of their products, it is a real-world effect that should not be underestimated.

All of this sheds a different light on the question that has been posed by a number of environmental film critics in the past years: whether or not a film that is designed to become a major blockbuster can approach an environmental topic seriously and affect the public in ways that are conducive to building healthier and more sustainable human-nature relationships. The staggering success of the latest mega-blockbuster with an openly environmental theme—James Cameron's *Avatar* (2009)—only makes this question more pressing. In this context, we should take seriously Lowe's contention in that "popular reporting of climate change in the style of environmental 'science fiction' ... creates a nagging concern, the solution to which is felt to be beyond the reach of the ordinary person" (2006b: 2). But this should not lead us—at least not directly—to the conclusion that popular film cannot function as "a catalyst for change" (2), as Lowe suggests. Rather, we should conclude that additional research needs to be conducted, ideally substantiated by audience response studies. Four of the five studies on *The Day After Tomorrow* were conducted in Western societies and all five in highly industrialized countries, which tells us nothing about receptions in countries like India and China, to name only two examples. As Anthony Leiserowitz puts it pointedly at the end of his article in *Environment*, "we have only scratched the surface ... in the effort to understand the role of popular representations of risk (such as movies, books, television, fiction, and nonfiction) or of cross-national differences in public risk perception and behavior" (44). A very good reason, I believe, to study such texts closely also from a cognitive perspective.

Works Cited

Aoyagi-Usui, Midori. "*The Day After Tomorrow*: A Study on the Impact of a Global Warming Movie on the Japanese Public." *National Institute for Environmental Studies (NIES). Working Paper* (unpublished), October 2004.

Balmford, Andrew, Percy FitzPatrick, Andrea Manica, Lesley Airey, Linda Birkin, Amy Oliver, and Judith Schleicher. "Hollywood, Climate Change, and the Public." *Science* 17 (September 2004): 1713.

Ban Ki-moon. Secretary-General's address to the IPCC upon the release of the Fourth Assessment Synthesis Report. 17 November 2007. Accessed 12 April 2009. <http://www.un.org/apps/sg/sgstats.asp?nid=2869>.

Beck, Ulrich. *World at Risk*. Cambridge: Polity Press, 2009.

Bordwell, David. "Cognitive Theory." *The Routledge Companion to Philosophy and Film*. London: Routledge, 2009. 356-67.

Bordwell, David and Noël Carroll (eds.). *Post-Theory: Reconstructing Film Studies*. Madison: University of Wisconsin Press, 1996.

Box Office Mojo. "Environmentalist." Accessed on 15 December 2010. <http://boxofficemojo.com/genres/chart/?id=environment.htm>.

———. "Controversy." Accessed on 15 December 2010. <http://boxofficemojo.com/genres/chart/?id=controversy.htm>.

Buell, Frederick. *From Apocalypse to Way of Life: Environmental Crisis in the American Century*. London: Routledge, 2004.

Busselle, Rick W. and Helena Bilandzic. "Emotion and Cognition in Filmic Narrative Comprehension and Engagement." Paper presented at the annual meeting of the International Communication Association, TBA, Montreal, Quebec, Canada, 22 May 2008. Accessed 17 December 2010. <http://www.allacademic.com/meta/p234265_index.html>.

Carroll, Noel. "Film, Emotion, and Genre." *Passionate Views: Film, Cognition, and Emotion*. Ed. Carl Plantinga and Greg M. Smith. Baltimore: Johns Hopkins University Press, 1999. 21-46.

Cubitt, Sean. *EcoMedia*. Amsterdam: Rodopi, 2005.

Damasio, Antonio. *Descartes' Error: Emotion, Reason, and the Human Brain*. 1994. London: Vintage, 2006.

The Day After Tomorrow. Dir. Roland Emmerich. Perf. Dennis Quaid, Jake Gyllenhaal. Twentieth Century Fox, 2004.

Hyde, William. "My Day After Tomorrow Review(ish) Thing." 4 July 2004. Accessed 5 July 2010. <http://groups.google.com/group/rec.arts.sf.written/msg/6e52157aaf63775f>.

Ingram, David. "Hollywood Cinema and Climate Change: *The Day After Tomorrow*." *Words on Water: Literary and Cultural Representations*. Ed. Maureen Devine and Christa Grewe-Volpp. Trier: Wissenschaftlicher Verlag Trier, 2008. 53-63.

Kakoudaki, Despina. "Spectacles of History: Race Relations, Melodrama, and the Science Fiction/Disaster Film." *Camera Obscura* 50 7.2 (2002): 109-53.

Keane, Stephen. *Disaster Films: The Cinema of Catastrophe*. London: Wallflower, 2001.

King, Geoff. *Spectacular Narratives: Hollywood in the Age of the Blockbuster*. London: I.B. Tauris, 2000.

Lang, Robert. *American Film Melodrama*. Princeton: Princeton University Press, 1989.

Leiserowitz, Anthony A. "Before and after *The Day After Tomorrow*: a U.S. Study of Climate Risk Perception. *Environment* 46.9 (2004): 23-37.

Lewis, David, Dennis Rodgers, and Michael Woolcock. "The Fiction of Development: Literary Representation as a Source of Authoritative Knowledge." *Journal of Development Studies* 44.2 (2008): 198-216.

Lowe, Thomas. "Is This Climate Porn? How Does Climate Change Communication Affect Our Perception and Behaviour?" *Tyndall Centre for Climate Change Research Working Paper 98* (December 2006a).

————. "Vicarious Experience vs. Scientific Information in Climate Change Risk Perception and Behaviour: a Case Study of Undergraduate Students in Norwich, UK. *Tyndall Centre for Climate Change Research Technical Report* 43 (May 2006b).

Lowe, Thomas, Katrina Brown, Suraje Dessai, Miguel de Franca Doria, Kat Haynes, and Katharine Vincent, "Does Tomorrow Ever Come? Disaster Narrative and Public Perceptions of Climate Change." *Tyndall Centre for Climate Change Research Working Paper* 72 (March 2005).

Mayer, Sylvia. "Teaching Hollywood Environmentalist Movies: The Example of *The Day after Tomorrow*." *Ecodidactic Perspectives on English Language, Literatures and Cultures*. Ed. Sylvia Mayer and Graham Wilson. Trier: Wissenschaftlicher Verlag Trier, 2006. 105-20.

Mercer, John and Martin Shingler. *Melodrama: Genre, Style, and Sensibility*. London: Wallflower, 2004.

Murray, Robin L. and Joseph K. Heumann. *Ecology and Popular Film: Cinema on the Edge*. Albany: SUNY Press, 2009.

Nisbet, Matthew. "Evaluating the Impact of *The Day After Tomorrow*: Can a Blockbuster Film Shape the Public's Understanding of a Science Controversy?" *CSICOP On-Line: Science and the Media*. 16 June 2004. Accessed 2 June 2010. <http://www.csicop.org/specialarticles/show/evaluating_the_impact_of_ the_day_after_tomorrow/>.

Plantinga, Carl. *Moving Viewers: American Film and the Spectator's Experience*. Berkeley: University of California Press, 2009a.

———. "Trauma, Pleasure, and Emotion in the Viewing of *Titanic*: A Cognitive Approach." *Film Theory and Contemporary Hollywood Movies*. Ed. Warren Buckland. AFI Film Readers. London: Routledge, 2009b. 237-56.

Plantinga, Carl and Greg M. Smith. "Introduction." *Passionate Views: Film, Cognition, and Emotion*. Ed. Carl Plantinga and Greg M. Smith. Baltimore: Johns Hopkins University Press, 1999. 1-17.

Reusswig, Fritz. "The International Impact of *The Day After Tomorrow*." *Environment* 46.9 (2004a): 41-43.

———. "Climate Change Goes Public: Five Studies Assess the Impact of 'The Day After Tomorrow' in Four Countries." *Potsdam Institute for Climate Research*. 5 November 2004. (2004b). Accessed 3 June 2010. <http://www.upf. edu/pcstacademy/_docs/200410_climatechange.pdf>.

Reusswig, Fritz, Julia Schwarzkopf, and Philipp Pohlenz. "Double Impact: The Climate Blockbuster *The Day After Tomorrow* and Its Impacts on the German Cinema Public." *PIK Report* 92 (October 2004).

Revkin, Andrew C. "When Manhattan Freezes Over." *New York Times* 23 May 2004. Accessed 12 June 2010. <http://query.nytimes.com/gst/fullpage.html? res=9503E5DB123FF930A15756C0A9629C8B63&sec=&spon=&pagewanted =3>.

Slovic, Paul. *The Perception of Risk*. London: Earthscan, 2000.

Slovic, Paul, Melissa L. Finucane, Ellen Peters, and Donald G. MacGregor. "Risk as Analysis and Risk as Feelings: Some Thoughts about Affect, Reason, Risk, and Rationality." *Risk Analysis* 24.2 (2004): 311-22.

Smith, Greg M. *Film Structure and the Emotion System*. Cambridge, MA: Cambridge University Press, 2003.

Smith, Jeff. "Movie Music as Moving Music: Emotion, Cognition, and the Film Score." *Passionate Views: Film, Cognition, and Emotion*. Ed. Carl Plantinga and Greg M. Smith. Baltimore: Johns Hopkins University, 1999. 146-67.

Staiger, Janet. *Perverse Spectators: The Practices of Film Reception*. New York and London: NYU Press, 2000.

Tan, Ed and Nico Frijda. "Sentiment in Film Viewing." *Passionate Views: Film, Cognition, and Emotion*. Ed. Carl Plantinga and Greg M. Smith. Baltimore: Johns Hopkins University Press, 1999. 48-64.

The Cuban Earthquake of 1880:
A Case Study from the Past with
Frightening Implications for the Future

Sherry Johnson

On 22 January 1880, the steamship *Admiral*, sailing from Cedar Key, Florida, arrived in Havana carrying a visiting party of North American dignitaries including former president Ulysses S. Grant, Lieutenant General Philip Sheridan, and their families. Barely twelve hours later, a strong earthquake struck the western half of the island of Cuba. Shocks were felt in Havana, but the greatest destruction occurred in the Vuelta Abajo region west of the capital city. Grant and his entourage were eyewitnesses to the quake, and reporters accompanying the party left sketches and newspaper articles as historical evidence of the event.

This chapter examines the Vuelta Abajo earthquake within the context of recent trends in disaster studies. Interdisciplinary theories about critical junctures, contingency, the nature of disasters, and the part they play as "triggers" for historical turning points are used to evaluate whether and to what extent the earthquake was a significant event in Cuban history. An examination of the circumstances surrounding the event will also suggest why it left no imprint on the historical memory of the island.

Theoretical Foundations

Critical juncture theory, incorporating the key issues of contingency, choice, and the ability to make a decision, are part of the cardinal theoretical framework used in this research. At every stage of a disaster—before, during, and after,—choices are made or not made that can be critical to the event's importance. According to critical juncture theory, a relatively minor event can escalate into a major catastrophe if its context is exacerbated by poor decisions taken by the major players (Capoccia and Keleman). A complementary question asks whether a natural hazard such as the Vuelta Abajo earthquake was actually a disaster or not (García Acosta 2010; Quarantelli)? More pointedly: Did the earthquake trigger a series of responses that led to changes in political, economic, and/or social behavior? (Olson and Gawronski 2007; Olson) Or, was it simply an event, a problem for the immediate victims that had no far-reaching consequences (García Acosta 2010; Olson and Gawronski 2007; S. Schwartz)? Another factor in determining the significance of a natural hazard is to evaluate its context—and the historical context of the Vuelta Abajo earthquake can show us why this disaster was quickly forgotten (Pfister et al.; Pfister). Finally, this chapter will show how this case study from the past can offer warnings for the present, particularly in light of the "environmental legacy of socialism" which is left by the course Cuba has taken over the past fifty years (Díaz-Briquets and Pérez-López).

The Geological, Geographical, and Historical Context

The series of faults that generated the earthquake were formed in the distant geological past when the Caribbean tectonic plate migrating northwestward collided with the West Atlantic plate. As the earth's crust pushed upward, the island of Cuba and the other islands of the West Indies became visible vestiges of the results of this collision. The more dangerous major fault line runs east to west well south of the island, but as it approaches the coast near the eastern province Oriente, it raises the potential for seismic activity in eastern Cuba, on the island of Hispaniola (as seen in the recent tragedy in Haiti) and in Puerto Rico (Calais et al.). A series of lesser faults criss-cross Cuba parallel to the main fault line, and among these, it was the virtually unknown disjuncture that underlies almost the entire western half of the island that shifted in January 1880 (Instituto Cubano 19).

The region west of Havana, the epicenter of the earthquake, is officially known as the province of Pinar del Río. Even today, the province has no major towns. Instead, the countryside is dotted with small villages dedicated to the region's major enterprise: tobacco farming. A line of hills runs east to west along the north coast while several streams—they can hardly be called rivers—originate in the hills and flow southward to the northern Caribbean sea. The valleys are punctuated by the limestone outcroppings characteristic of karst topography. Near the southern coast, the terrain flattens into a broad alluvial plain where the fertile soil along the margins of the streams produces the exquisite Cuban tobacco so prized by cigar aficionados worldwide (Instituto Cubano 19).

If the context of a disaster can influence whether or not it becomes a critical juncture, the Vuelta Abajo earthquake had the potential to be such a moment, given the historical setting in which it occurred. From 1868 through 1878, the island was engulfed by a rebellion that hoped to achieve independence from Spain for its inhabitants but resulted in little more than extensive property destruction, the economic ruin of many of the primary families on the island, and even greater political conflict. In October 1868, a group of influential men in the eastern region launched a revolt against Spanish rule (Ferrer). For a whole decade (which lent the conflict its name, the "Ten Years War"), the insurrection tore the island apart. The eastern part of the island was largely in favor of independence, while in the western half most people acquiesced to Spanish rule. Even if they were politically neutral, many families were ruined by the fighting and by reprisals against them for not taking sides (Quiroz 1998). The war was also responsible for the first sustained wave of exiles who fled to the United States.

After the armistice in 1878, in the form of the Pact of Zanjón, which accomplished little more than reinstating the *status quo ante bellum*, Spain promoted emigration from the peninsula (Ferrer; Casanovas). The sole positive aspect of the armistice was a provision that led to the abolition of African slavery in Cuba, implementing an apprenticeship system under which any person still enslaved in 1878 would be freed by 1888 (Scott). The Ten Years' War in Cuba combined with civil war in Spain left both mother country and colony exhausted and in financial ruin, and Spain had little choice but to encourage large-scale foreign investment in Cuba (Quiróz 2003; Fernández; *Brooklyn Eagle* hereafter *BE*, 6 Feb 1880). Railroad construction, which had begun in the 1830s, now took a giant leap forward with Canadian companies, helped by significant US investments, leading the way (Santamaria and García; Zanetti and García). Railroads and steamship lines brought hordes of tourists to Cuba via Florida who demanded further improvements in the infrastructure to maintain their comfort and pleasure (*Diario de la Marina*, Havana, hereafter *DM*, 9 Jan 1880). The demands of the tourists could be met only with difficulty; if the political, economic, and social turmoil of the previous decade was not enough, two years earlier a devastating hurricane had added to the island's misery, compounding the residual problems left over from the war (Archivo Histórico de la Nación, hereafter AHN).

The Vuelta Abajo Earthquake of 1880

Around 11:30 p.m. on 22 January 1880, the first shock waves rolled across Vuelta Abajo followed by a stronger wave of oscillations around 4:00 a.m. (*New York Times,* hereafter *NYT*, 25 Jan 1880). In Havana, already known as the playground of the Caribbean, nightly entertainment was in full swing (R. Schwartz). The casinos, brothels, and taverns were crowded with people from many different countries enjoying the fabled pleasures of the Cuban capital. Some less hearty visitors had already retired for the night when the first tremors hit the city. Hotel guests were jolted awake by the floors in their rooms swaying and buckling, cracks opening in the walls, and plaster raining down upon them. Survival instincts took over as terrified men and women rushed out of their hotels into Havana's many public squares. The majority were only partially clad in bed sheets or whatever they could grab to cover themselves, while more than a few were clothed in "little but the moonlight"(*NYT* 28 Jan 1880). In bars and gaming halls, bottles and glasses came crashing to the floor, interrupting the nightly work of Havana's prostitutes and pickpockets as their clientele abandoned their merrymaking and joined the crowds in the streets. Fright, disorientation, various stages of inebriation, and a multitude of languages added to the confusion. One old reprobate knelt down to pray for his life, but he could only remember the first line of "Now I lay me down to sleep" (Chicago *Inter-American*, quoted in the *NYT* 28 Jan 1880). President Grant and his entourage were lodged in the captain general's palace, a sturdy building constructed in the late eighteenth century to house the ranking political and military officials in Cuba. According to the on-site correspondent for the Chicago *Inter-American*, the president and his group were unharmed, although the ladies in the party "were terribly frightened." The reporter assured his readers that Grant and General Sheridan "were not disturbed in the least" (*NYT* 28 Jan 1880). Daylight brought more aftershocks along with the good news that for the most part, Havana had escaped the worst of the destruction.

Image 1: *Recollections of a Cuban Earthquake*. Frank Hamilton Taylor.
Sketches of Cuba, from the pamphlet "A Stately, Picturesque Dream"

The tremors from the earthquake were felt along the north coast of the island, from the bay of Mariel to Matanzas on the north coast and as far away as Cienfuegos to the southeast, a distance of approximately two hundred miles, or three hundred kilometers (AHN). While the initial reports stated that the earthquake had originated in the Florida Keys or the Bahamas, it became apparent that the epicenter was on land west of the capital (*NYT* 28 Jan 1880; AHN). The town closest to the epicenter, San Cristóbal, was the hardest hit. Early reports informed that the church, the town hall, the jail, the guardhouse, and the telegraph office, all made of mortar and stone, were destroyed. Houses of similar construction collapsed while, ironically, wooden houses and the wattle-and-daub huts of the tobacco farmers survived the quake with minor damage. Eyewitnesses reported that after the strong shock at 4:30 in the morning, all of the clocks instantly stopped (*NYT* 6 Feb 1880; *BE* 6 Feb 1880). In the neighboring village, Candelaria, the mortar and stone church collapsed and other buildings were damaged, while in the village of San Diego de los Baños, known for its hot springs, the water in the mineral baths became cloudy and unfit for bathing. No lives were lost in Candelaria, San Diego, or San Cristóbal, but seventeen rural guards were injured, three seriously, prompting the government authorities to dispatch a corps of military engineers to survey the damage, to take over the police functions of the rural guard, and to prevent looting. Northward, across the line of hills near Cabañas Bay, one death was reported on a small sugar plantation, San Juan Bautista, when the boiling house collapsed around one worker, killing him instantly and seriously injuring his wife along with several men in his harvesting crew. The regional governor of the province rushed from his administrative office in Pinar del Río to San Cristóbal to supervise the situation and to inform the captain general in Havana of what was needed to recover from the disaster. The captain general promised to "respond with every resource that this government has to offer" (AHN).

For several days thereafter, the region experienced a series of aftershocks accompanied by a rumbling sound coming from deep within the earth (*NYT* 6 Feb 1880; *BE* 6 Feb 1880). In some areas, cracks opened in the ground, and toxic sulfurous gas poured into the air. Fissures over a foot wide appeared in the tobacco fields located in the bottomlands of the numerous rivers, while stones, rocks, and fossils carried by water from the subterranean layers poured out over the tobacco crop (*NYT* 6 Feb 1880; *BE* 6 Feb 1880). More tremors rolled across the region throughout the spring and into the summer of 1880, and as late as April of the following year, the ground under western Cuba continued to shake as the fault line adjusted in its movement northwestward (*NYT* 22 Aug 1880, 6 Apr 1881).

Unlike earlier times, by the late nineteenth century there was no shortage of reports about the earthquake, especially since former president Grant and his entourage had arrived in Cuba just that morning (see image 2 on next page). A correspondent and sketch artist for *Harper's Weekly*, Frank Hamilton Taylor, accompanied the Grant party as they travelled to the epicenter, San Cristóbal, to view the damage (Gustke). Taylor's pen-and ink drawings of the conditions in San Cristóbal and Havana provide visual evidence of the aftermath.

Image 2: *Grant's arrival.* Frank Hamilton Taylor.
Sketches of Cuba, from the pamphlet "A Stately, Picturesque Dream"

Over the course of several months, his sketches were turned into woodcuttings and later into engravings that were published in *Mechanical News* in 1886. In addition, by 1880, several international newspapers had correspondents in Cuba, and the reports they sent home to their publishers arrived quickly thanks to the new transatlantic telegraph cable that linked Cuba to the outside world. When the earthquake struck, the correspondents were the first to be in the plazas gathering information from the frightened tourists and residents. Other eyewitness accounts came from workers on the Eastern Railroad that was under construction in Vuelta Abajo and from local officials in the region (*NYT* 6 Feb 1880; *BE* 6 Feb 1880). By all accounts, though, public opinion was unanimous in agreeing that the Vuelta Abajo earthquake took everybody by surprise. Even the most celebrated scientist of the day, Jesuit priest Fr. Benito Viñes of the Observatory of Belén in Havana (known for his work on hurricanes), was stunned that the western region of the island could be subject to earthquakes, as there was no mention of any earthquake ever hitting the region in the records that had been kept since the western part of the island was settled in 1519 (AHN).

Image 3: Collapsed porch on one house, but note the surrounding wooden houses appear to show little damage. *After the Earthquake.* Frank Hamilton Taylor. Sketches of Cuba, from the pamphlet "A Stately, Picturesque Dream"

Image 4: The cracks in the walls of the mortar structure made the entire building unstable. *The Prison after the Earthquake.* Frank Hamilton Taylor. Sketches of Cuba, from the pamphlet "A Stately, Picturesque Dream"

Image 5: Soldiers on the plaza keeping order. They seem to be making no effort
to help civilian victims. *Plaza after the Earthquake*. Frank Hamilton Taylor.
Sketches of Cuba, from the pamphlet "A Stately, Picturesque Dream"

Image 6: Victims coping with the aftermath. The family outside their ruined
home is cooking a meal. *A Meal after the Earthquake*. Frank Hamilton
Taylor. Sketches of Cuba, from the pamphlet "A Stately, Picturesque Dream"

In late January, Grant and his entourage left Havana to visit the mineral baths in
San Diego. They planned to make the journey by railroad, but they were forced to
transfer to a carriage, the characteristic transport of the Cuban countryside, because
the earthquake had compromised the integrity of the bridges that spanned Vuelta

Abajo's many streams. Yet, in spite of the reports, the visual evidence, and Viñes' opinion that "it was an event unprecedented in the history of the island," the Vuelta Abajo earthquake had little impact beyond its immediate area (AHN). The flagship newspaper of the Americas, the *New York Times,* initially ran several stories, but in just a few short weeks, the event was all but forgotten. The last accounts of the earthquake appeared in United States newspapers on 6 February, contemporaneous with the former president's return to Havana. Unfazed by the earthquake, he and his party continued their grand tour by travelling eastward to Matanzas, where they visited local attractions such as the famed cave of Bellamar. By 15 February, Grant was back in Havana enjoying the sights and sounds of carnival, the pre-Lent celebration of excess common in Catholic countries (*DM* 15 Feb 1880), and by the end of the month, Grant and his companions had left Cuba and were en route to Mexico. The earthquake passed virtually unnoticed into the historical record, and even today, few *piñarenos* or *habaneros* realize that they live atop a major fault line that could generate another earthquake at any time.

Comparing, Contrasting, and Integrating Theoretical Approaches to Disaster

Until Hurricane Andrew devastated South Florida in 1992, catastrophe received scant attention as a conceptual tool to establish and to evaluate historical processes. Andrew brought home to scholars the effect of a disaster and its aftermath in a most painful fashion. In the context of this study, earthquakes as catalytic events in and of themselves also contribute to a growing body of research that takes disasters as its starting point and examines the consequences of such events (cf. Buchenau and Johnson; Dauer; Mulcahy; Walker; Winder). Combined with works that establish the theoretical foundations of disaster studies, this scholarship permits an evaluation of the importance of the Cuban earthquake of 1880. For example, seminal works in political science demonstrate that disaster can be a force behind political change, but disasters do not necessarily have to become political (Drury and Olson; García Acosta 1996; Olson). The authorities' behavior in the aftermath of disaster determines whether the population will react in a positive or a negative way, thus making the disaster the trigger that causes a "critical juncture" in political events (Olson and Gawronski 2007). The concepts implicit in critical juncture theory rest upon the idea of contingency, that is, acknowledging that many potential paths could be chosen, most of which would lead to a different outcome—some positive, some negative (Capoccia and Keleman). Anthropology and sociology, especially studies of the social chaos after a crisis event, promise to lend yet additional conceptual tools (Kreps; Oliver-Smith; Peacock et al.). Especially useful are anthropological and sociological studies that examine the leveling effect of disaster, the post disaster community self-organizing efforts, and their innate ability to recover ("resilience") (Pérez). Although social boundaries would be restored as life returned to normal, one's behavior during the emergency would remain in the community's collective memory (Provenzo and Baker Provenzo). Bravery and decisive positive decisions were embedded in the community's memory in the form of folk tales and stories, while cowardice, ineffectualness, and obstructionism would also be remembered by the stricken residents (Johnson).

Using historical documentation, scientific evidence, and interdisciplinary theory to explain historical processes is undeniably seductive, yet the temptation to attribute change over time to a catastrophe of any sort must be tempered with common sense (Olson and Gawronski 2007; S. Schwartz). As Virginia García Acosta has pointed out, not every natural hazard turns into a disaster (2010). To avoid the fallacy of assigning too much significance to disasters, this cautionary literature asks whether the catastrophe produced "legacies," that is, if permanent change over time can be attributed wholly or in part to the event (Olson and Gawronski 2007).

The Vuelta Abajo earthquake was neither a critical juncture nor a trigger that changed government policy in any way. Except for one fatality, there were no casualties, minimal property damage, and no famine or food shortages as the earthquake occurred in the winter when there is usually plenty of food available. In short, there was no reason for Cuban people to remember it. Instead, it was, as Olson and Gawronski have pointed out, a "null event," that is, a natural phenomenon that comes and goes just as quickly with few, if any, consequences (2007).

The real importance lies in the implications for the present of lessons *not learned* about the seismic potential of western Cuba; as such, the unprecedented earthquake of 1880 should be taken as a warning by the residents of the region. The fault line still lies underneath the cities and towns that house a majority of the Cuban population, and it continues to migrate northwestward at the rate of eleven centimeters per year. Such an observation is alarming when combined with an awareness of the "environmental legacy of socialism" in Cuba (Díaz-Briquets and Pérez-López). In the previous fifty years, the island has suffered deterioration in every aspect of its infrastructure: roads, bridges, wharves, buildings, homes, hospitals, and so on. Havana and its suburbs are notorious for their crumbling appearance and substandard living conditions. Even a minor summer storm can contribute to the collapse of a building. Torrential rains and high winds work together to abrade ancient mortar, and as pieces of deteriorating buildings fall to the ground, they sometimes pierce the exposed antiquated natural gas lines, which were put in place around 1900 during the first US occupation of the island. Spontaneous fires are sometimes the result (observed by the author personally in 1992, 1993, 1995). An antiquated potable water delivery network and an inadequate sewer system exacerbate daily living conditions for Havana's residents. Even a minor earthquake could create a post-disaster situation much like that which occurred in Port au

Prince in January 2009. Cuban residents who might escape the destruction caused by the initial tremors would undoubtedly be faced with the sanitation crisis that ensued after the Haitian quake.

But even the massive loss of life that would occur should another earthquake happen along this fault line pales in comparison to the threat posed by a second aspect of socialism's environmental legacy: the inadequate protection of nuclear materials. The line of hills south of Havana, formed by the uplifting of the land in the geological past, provides an ideal environment to store leftover nuclear waste from a variety of sources such as medical applications and military experimentation. The storage facility near the town of Managua, built during the time of cooperation with the Soviet Union, lacks the requisite safety features to contain escaping radiation in the event of a breach. A similar nuclear nightmare is possible if the epicenter of the next quake moves just one hundred kilometers to the east to Juraguá on the Bay of Cienfuegos, where an unfinished nuclear plant built with Soviet technology was ostensibly never activated. According to Cuban sources, the potential for disasters was taken into account when the plant was under construction, but at the time, engineers were thinking in terms of hurricanes. No provision was made for potential seismic activity (Díaz-Briquets and Pérez-López). Moreover, the Cuban government maintains that no nuclear material was ever brought to the plant, but no outside inspectors have ever been allowed to visit. If no nuclear material is present in the Juraguá plant, and if the fault line generates another earthquake, the only consequences would be to the residents of the closest city, Cienfuegos. But if any dangerous material is actually housed in the defunct nuclear plant, the loss of life could be unimaginable.

The residents of western Cuba, thus, face a potential catastrophe that combines the consequences of the Port-au-Prince earthquake of 2009 and the nuclear meltdown at the Fukushima Dai-ichi nuclear plant in Japan in 2011. Both scenarios, equally frightening, are the inescapable lessons that could be learned from the Vuelta Abajo earthquake of 1880. Critical juncture theory explains to us why this event was forgotten, lost to memory; and thus it is up to historians to tell these stories, both of iconic events such as the Lisbon earthquake and the San Francisco fire, and the forgotten disasters such as this, the Vuelta Abajo earthquake, in the hope that a culture of reintegrating our knowledge of catastrophes past might yet save us from the worst conjunctions of human forgetfulness and natural disaster.

Works Cited

Primary Sources

AHN, Archivo Histórico de la Nación, Madrid, Spain. Ultramar, legajo 219, expediente 13, January 25, 1880. Accessed via Portal de Archivos Españoles [PARES] <www.pares.mcu.es>.

Taylor, Frank H. Watercolor and Engraving Collection. Museum collection, University Gallery purchase, Samuel P. Harn Museum of Art, University of Florida, Gainesville.

New York Times 25 January, 28 January, 6 February, 25 February, 22 August 1880; 6 April 1881.

Brooklyn Eagle 11 January, 6 February 1880.

Diario de la Marina (Havana). 9 January, 15 February 1880.

Secondary Sources

Buchenau, Jürgen and Lyman L. Johnson (eds.). *Aftershocks: Earthquakes and Popular Politics in Latin America*. Albuquerque: University of New Mexico Press, 2009.

Calais, E., J. Perrot, and B. Mercier de Lépinay. "Strike-Slip Tectonics and Seismicity along the Northern Caribbean Plate Boundary from Cuba to Hispaniola." *Active Strike-Slip and Collisional Tectonics of the Northern Caribbean Plate Boundary Zone*. Ed. James F. Dolan and Paul Mann. Boulder: Geological Society of America, 1998.

Cappocia, Giovanni and R. Daniel Keleman. "The Study of Critical Junctures: Theory, Narrative, and Counterfactuals in Historical Institutionalism." *World Politics* 59.3 (2007): 341-69.

Casanovas, Joan. *Bread, or Bullets!: Urban Labor and Spanish Colonialism in Cuba, 1850-1898*. Pittsburgh: University of Pittsburgh Press, 1998.

Dauer, Quinn. "Bitter Wine: The Mendoza Earthquake of 1861 and the Formation of the Argentine State." Paper presented at the American Historical Association, Conference on Latin American History Joint Session, American Historical Association Meeting, San Diego, CA, January 2010.

Diaz-Briquets, Sergio and Jorge F. Pérez-López. *Conquering Nature: The Environmental Legacy of Socialism in Cuba*. Pittsburgh: University of Pittsburgh Press, 2000.

Drury, A. Cooper and Richard Stuart Olson. "Disasters and Political Unrest: An Empirical Investigation." *Journal of Contingencies and Crisis Management* 6.5 (September 1998): 153-61.

García Acosta, Virginia. "Social Disasters in Mexico: Vulnerabilities, Risks, and Adaptive Strategies." Paper presented at the Rachel Carson Center Lunchtime Colloquium, 25 June 2010.

———. "Introduction." *Historia y desastres en América Latina.* Vol. 1. Coord. Virginia García Acosta. Mexico City: Centro de Investigaciones y Estudios Superiores en Antropología Social (CIESAS), 1996.

Gustke, Nancy L. "Introduction." *"A Stately Picturesque Dream ...": Scenes of Florida, Cuba, and Mexico in 1880; Forty-Seven Brush Drawings and Watercolors.* Gainesville: University Gallery, College of Fine Arts, University of Florida, 1984.

Fernández, Susan J. *Encumbered Cuba: Capital Markets and Revolt, 1878-1895.* Gainesville: University Press of Florida, 2002.

Ferrer, Ada. *Insurgent Cuba: Race, Nation, and Revolution, 1868-1898.* Chapel Hill: University of North Carolina Press, 1999.

Instituto Cubano de Geodesia y Cartografía. *Atlas de Cuba: XX aniversario del triunfo de la revolución cubana.* La Habana: Instituto Cubano de Geodesia y Cartografía, 1978.

Johnson, Sherry. *Climate and Catastrophe in Cuba and the Atlantic World in the Age of Revolution.* Chapel Hill: University of North Carolina Press, 2011.

Kreps, Gary A. (ed.). *Social Structure and Disaster.* Newark: University of Delaware Press, 1989.

———. "Sociological Inquiry and Disaster Research." *Annual Review of Sociology* 10 (1984): 309-30.

Mauch, Christof and Christian Pfister (eds.). *Natural Disasters, Cultural Responses: Case Studies toward a Global Environmental History.* Lanham: Lexington Books, 2009.

Mulcahy, Matthew. "The Rise and Fall—and Rise and Fall Again—of Port Royal, Jamaica." Paper presented at the American Historical Association/Conference on Latin American History Joint Session, American Historical Association Meeting, San Diego, CA, January 2010.

Oliver-Smith, Anthony. "Anthropological Research on Hazards and Disasters." *Annual Review of Anthropology* 25 (1996): 303-28.

Olson, Richard Stuart. "Towards a Politics of Disaster: Losses, Values, Agendas, and Blame." *International Journal of Mass Emergencies and Disasters* 18.2 (August 2000): 265-87.

Olson, Richard Stuart and Vincent T. Gawronski. "From Disaster Event to Political Crisis: A '5C+A' Framework for Analysis." *International Studies Perspectives* 11.3 (August 2010): 1-17.

———. "Disasters as Crisis Triggers for National Critical Junctures? The 1976 Guatemala Case." Paper presented at the Southeastern Council of Latin American Studies Meeting, San Jose, Costa Rica, 2007.

Peacock, Walter Gillis, Betty Hearn Morrow, and Hugh Gladwin (eds.). *Hurricane Andrew: Ethnicity, Gender, and the Sociology of Disasters*. London: Routledge, 1997.

Pérez, Jr., Louis A. *Winds of Change: Hurricanes and the Transformation of Nineteenth-Century Cuba*. Chapel Hill: University of North Carolina Press, 2001.

Pfister, Christian. "Learning from Nature-Induced Disasters: Theoretical Considerations and Case Studies from Western Europe." *Natural Disasters, Cultural Responses: Case Studies Toward a Global Environmental History*. Lanham: Lexington Books, 2009. 17-40.

Pfister, Christian, Emmanuel Garnier, Maria Joao Alcoforado, Dennis Wheeler, Jurg Lauterbacher, Maria Fatima Nunes, and Joao Paolo Taborda. "The Meterorological Framework and the Cultural Memory of Three Severe Winter Storms in Eighteenth-Century Europe." *Climatic Change* 101.1-2 (2010): 281-310.

Provenzo, Jr., Eugene F. and Asterie Baker Provenzo. *In the Eye of Hurricane Andrew*. Gainesville: University Press of Florida, 2002.

Quarantelli, E.L. (ed.). *What Is a Disaster: Perspective on the Question*. London: Routledge, 1998.

Quiroz, Alfonso W. "Implicit Costs of Empire: Bureaucratic Corruption in Nineteenth-Century Cuba." *Journal of Latin American Studies* 35.3 (August 2003): 473-511.

———. "Loyalist Overkill: The Socioeconomic Costs of 'Repressing' the Separatist Insurrection in Cuba, 1868-1878." *Hispanic American Historical Review* 78.2 (May 1998): 261-305.

Santamaría García, Antonio and Alejandro García Alvarez. *Economía y colonia: la economía cubana y la relación con España, 1765-1902*. Madrid: Consejo superior de investigaciones científicas, 2004.

Schwartz, Rosalie. *Pleasure Island: Tourism and Temptation in Cuba*. Lincoln: University of Nebraska Press, 1999.

Schwartz, Stuart. "Hurricanes and the Shaping of Circum-Caribbean Cultures." Keynote address at the Third Biennial Allen Morris Conference on the History of Florida and the Atlantic World, Tallahassee, FL, 2004.

Scott, Rebecca J. *Slave Emancipation in Cuba: The Transition to Free Labor, 1860-1899*. Princeton: Princeton University Press, 1985.

Taylor, Frank H. *"A Stately Picturesque Dream ...": Scenes of Florida, Cuba, and Mexico in 1880: Forty-Seven Brush Drawings and Watercolors*. Gainesville: University Gallery, College of Fine Arts, University of Florida, 1984.

Walker, Charles F. *Shaky Colonialism: The 1746 Earthquake-Tsunami in Lima, Peru, and Its Long Aftermath*. Durham: Duke University Press, 2008.

Winder, Gordon. "Mediating Foreign Disasters: *The Los Angeles* Times and International Relief, 1891-1914." *Historical Disasters in Context: Science, Religion, and Politics*. Ed. Andrea Janku, Gerrit Schenk, and Franz Mauelshagen. New York: Routledge, forthcoming.

Zanetti, Oscar and Alejandro García. *Sugar & Railroads: A Cuban History, 1837-1959.* Translated by Franklin W. Knight and Mary Todd. Chapel Hill: University of North Carolina Press, 1998.

The *Los Angeles Times* Reports Japanese Earthquakes, 1923-1995[1]

Gordon Winder

Content to cover foreign disasters by printing bulletins off the wire from Associated Press (AP) for its first two decades, *The Los Angeles Times* (*LAT*) covered its first in-house earthquake disaster, including extensive reports of charity initiatives in Los Angeles, on the San Francisco earthquake of 1906 (Winder). Four months later, following another major earthquake in Valparaiso, Chile, on 18 August 1906, the *LAT* transferred *this* framing and practice to its reports of an international earthquake for the first time. Thus, in this newspaper, a pattern for media event reporting of distant earthquake disasters was set in 1906. Since then the *LAT* has generally been in step with other US media, which have been fascinated by foreign disasters, and have built upon a strong national humanitarian impulse to turn foreign disasters into aid campaigns, in the process making charity a high profile practice in the public sphere (Rosenblum; Hallin; Benthall; Rotberg and Weiss; Winder; Moeller; Hutchinson). In each case, the media conveyed a moral tale of disaster but not one framed by issues of preparedness. Science stories surfaced but were ancillary to the main form of reader engagement: a media ritual of Americans organizing relief and expressing sympathy for distant victims. The media uses a 'journalism of exception' (Lynch et al. 269-312) in which domestic news is portrayed as distinct from international news which is full of war, conflict, and disaster. Myths of catastrophe furnish the language and the key themes of modern US disaster reporting (Bird et al. 341; Heyer). The US media writes the international disaster story as a morality tale designed to affirm and uphold the existing social order in the United States, and to "reaffirm US political authority and superiority on the world stage" (Lule 183).

However, little is known about the changing ways in which subsequent twentieth century earthquakes were reported within this mainstream, metropolitan newspaper, or about the changing ways in which readers were situated in these stories in

[1] Research for this conference paper was partly funded by a University of Auckland research grant. This paper was presented at the Japan Studies Institute, Ludwig-Maxmilians Universität, Munich, and benefitted from the discussion there. The present chapter has benefited from the advice of *The Canadian Geographer* reviewers and editors Roger Hayter and Ian MacLaughlan. Finally, I also thank the staff, students and research fellows at the Rachel Carson Center, LMU Munich for their interest and comments concerning this research.

terms of moral tales and geographical imaginaries about the world, news production, and relations with others (Cosgrove et al.; Hannah). This chapter examines
the *LAT's* reporting of nine devastating Japanese earthquakes that occurred at
widely distributed dates over the seventy-two years from 1923 to 1995.

Table 1: Major Japanese Earthquakes 1891-1995

Location	Date	Year	Magnitude (R)	Deaths (No.)	LAT Coverage		
					Words (No.)	AP (%)	Japan (%)
Mino-Owari	October 27	1891	8.0	7,273	731	92.3	10.2
Sanriku	June 15	1896	8.5	27,000	0	0	0
Kanto	September 1	1923	7.9	143,000	76,876	28.5	4.0
Tango	March 7	1927	7.6	3,020	4,222	77.5	67.5
Sanriku	March 2	1933	8.4	2,990	1,025	64.4	100.0
Tottori	September 10	1943	7.4	11,990	309	62.5	0
Tonankai	December 7	1944	8.1	1,223	1,302	54.8	0
Mikawa	January 12	1945	7.1	1,961	0	0	0
Nankaido	December 20	1946	8.1	1,330	4,792	82.1	81.5
Fukui	June 28	1948	7.3	3,769	2,051	49.4*	84.8
Kobe	January 16	1995	6.9	5,502	52,697	2.5	12.5

Note: *Together AP and UPI contributed 93.4%; source: US Geological Survey

Analysis of the *LAT* reports of these events reveals how the newspaper's editors
and journalists changed their narrative and their news gathering practices. This
chapter asks how did the *LAT* frame Japanese earthquakes as stories relevant to
Los Angeles readers, and how did this framing change through the twentieth century as globalization processes altered the environment for newspapers? The analysis focuses on how news of foreign disasters constructs imagined places, communities, and identities. My research asks: did the geographical imaginaries and
relations in the earthquake news change, and if so, how? How were the imagined
readers and their possible engagement with the story recast? And how have the
identities and relations ascribed to victims, rescuers, scientific knowledge, communications, and nations been reworked?

   This chapter builds upon the ideas of imagined communities, media events, and
media rituals (Anderson; Dayan et al.; Couldry; Cottle) by directing attention to
how place is constructed within news stories. This opens the way for a history of
disaster reporting that reflects the USA's changing foreign relations. The *LAT* generates and circulates stories about earthquake disasters using geographical imaginaries of the world and 'citizenship' within it for its readership in Southern California. Specific values are credited to geographical imaginaries: for example the
USA is portrayed as a wealthy and generous donor; or the Japanese government is
said to have failed to protect its citizens. In these ways newspaper disaster reports
should reflect (if not generate) geographical learning. In addition to accounting the

scale of the disaster in terms of a human death toll and the devastation of land-scapes, the *LAT* presents readers with stories that situate them in (geographical) re-lations with communications technology, scientific experts, teams of fund-raisers and rescuers, politicians, state authorities, victims, and survivors. These relation-ships are constituted in terms of drama, ritual, and morality. The readers' engage-ment with distant victims is ritualized and dramatized as an account of their gener-osity as charitable aid donors. Even the work of producing the story is made into a narrative in space as the newspaper informs readers of where the story is compiled, under which conditions, and how reliable or authoritative it is. Thus the modern disaster news story has a narrative structure that constitutes readers in the world, and affirms the existing social order. As distant places become more well known to readers, and as readers' and journalists' 'ideoscapes' (Appadurai 27-47) and geo-graphical imaginations are challenged and reconstituted, so the elements of the story should be recast.

    We should find an identifiable development of the modern earthquake disaster story within the pages of the *LAT* as globalization processes take effect (Wilke). In the course of the last century, new telecommunications technology altered the rela-tionships between readers, journalists, editors, and participants, eventually making newspapers a less dominant news media, as first radio and then television captured news markets. Newspapers developed their own news production capabilities, in-cluding foreign bureaux, and thus reduced their dependence on news agencies. In the 1990s, Los Angeles was a shock city of post-modernity (Davis; Soja; Scott et al.), densely networked into the global economy as a global city, and this was par-alleled in the way the *LAT* reported the Kobe disaster. Now the information tech-nology associated with the internet and cell phones is democratizing the public sphere. Thus, in early 2011, the world-wide circulation of images of the smoke hanging over the disabled Fukushima nuclear plant and the debris left by the earth-quake and tsunami in northern Japan triggered anti-nuclear protests in many coun-tries. However, other influences on the newspaper's disaster reporting may have cut across the effects of the globalization dynamic and this paper considers two ad-ditional influences. It may be that the *LAT* reveals a specific US culture of disaster, even when it reports foreign earthquakes (Bankoff[2]). Further, the USA's diplo-matic relations with foreign powers may have been important in framing disaster reporting.

    US media coverage of disasters is narrowly framed. Environmental historian Theodore Steinberg contends that in the USA "modern disaster discourse" is so

---

[2]   Environmental historian Greg Bankoff has identified national cultures of disaster that include ways of coping with repeated disasters.

thoroughly adopted that nature is perceived as the villain: "natural disasters have come to be seen as random, morally inert phenomena" (Steinberg xxii). For Steinberg this means that ethical responsibility and ecological literacy are hard to pin down in a nation where everyone (and no one) bears the cost of living in a hazard zone. But Steinberg's modern disaster discourse denies the other modern claim, made by geographer Ken Hewitt, that there are no 'natural disasters,' only vulnerable populations needing expert planning (Hewitt; Bolin et al.; Oliver-Smith). In this context, Birkland found that one third of the 345 stories on earthquakes published in *The New York Times* 1990-2002 reported the size of the event and a quarter noted the damage done; science, future threats and recovery were common features of news stories, but mitigation, preparedness, relief and response—the four stages of disaster response—were barely mentioned (Quarantelli 1988; Quarantelli 1997; Birkland 117). In these ways social scientists have charted a specific US disaster discourse appropriate to earthquakes. It features charitable aid, minimizes mitigation, preparedness, and response, confuses and occludes responsibility, and asserts US superiority among nations.

American sympathy for Japanese victims of earthquake disasters was tempered by the vicissitudes of the troubled relationship between the two countries. In the last decade of the nineteenth century and the first two decades of the twentieth, Japan tended to be perceived in the United States as a modernizing Asian nation, and as a partner in maintaining the USA's Open Door policy in China. The relationship soured in the 1920s, when Japan was increasingly seen as an imperial rival, indeed one aggressively pursuing expansionist interests in China. How could the *LAT* solicit US aid for Japanese earthquake victims when the two countries were at war or when US military forces occupied the Japanese archipelago? The US-Japan relationship improved during the Cold War, and the current period of globalization constitutes a new geopolitical context for reports of Japanese earthquakes in the US media. The history of the US-Japan relationship cuts across the effects on the disaster reporting of globalization and modernization processes, but the precise effects remain to be analyzed.

This chapter compares and contrasts the presentation of earthquake stories across time in the *LAT*. The coverage of those Japanese earthquakes which killed more than one thousand people is analyzed (Table 1). The 1891 Mino-Owari and 1896 San Riku earthquakes occurred before the *LAT* began to offer extensive coverage of foreign disasters. For these earthquakes it offered either no report or off-the-wire copy from AP. However, the *LAT*'s reports of the 1923 Tokyo or Kanto Plain earthquake show the intensification of the general patterns of engagement between California and the disaster zone that is apparent in its reports of foreign earthquake disasters following the San Francisco and Valparaiso earthquakes of 1906 (Winder). Notably, the economic impacts of the disaster on the USA, the fates of Americans located in the disaster zone, and the aid campaigns launched by

Californians and the American government are reported at length. In the week after the event, the *LAT* offered its readers 77,000 words on the Kanto Plain earthquake. Only four percent of this news was actually sourced in Japan, and AP was no longer as important a source as it had been. Coverage of the earthquakes of Japan's Depression era, wartime, and Cold War earthquakes (Table 1) show a return to the earlier format: the *LAT* offered much reduced coverage and relied upon AP copy, sourced in Japan. In this context, the *LAT*'s reporting of the Kobe earthquake of 1995 is more like its coverage of the 1923 Tokyo quake than the other earthquakes listed here. Not only did the *LAT* print over 52,000 words on this event, but now seventy percent of this copy was from Los Angeles. The newspaper highlighted controversy over the extent and character of Washington's role in the relief work and in reconstruction.

So, while each of these earthquakes was a significant natural disaster in terms of both magnitude and death toll, two earthquakes stand out, and necessarily, the chapter focuses on what was published about natural disasters, Japan, and Americans in the immediate aftermath of the terrible earthquakes that devastated Tokyo in September 1923 and Kobe in January 1995. By focusing on coverage in one US newspaper of disasters in one other country spread over a long period of time, this chapter is able to discern patterns of continuity and change in the stories that are told. Thus, analysis of coverage of these events outlines a general trajectory of development within the *LAT*'s earthquake disaster story.

No attempt is made here to study the life history of each disaster, though this is a feature of research on the political ecologies of disaster (Oliver-Smith et al. 30). Instead, copies of the *LAT* (Table 1) were collected for the week following the event.[3] Earthquake coverage in these 'news weeks' was then mapped and analyzed using spreadsheets of authors, agencies, datelines, headlines, word counts and images. This approach permits a reading of the changing geographies of news production. The narrative structures of the stories told are analyzed along with the themes and moral messages written into these news reports. The 'whole newspaper' approach followed here also facilitates appreciation of the broader context for disaster news. The roles of gatekeepers in making decisions about what becomes news cannot be discerned in this approach and methodology, but the general framing of the news can (Whitney et al.; Singer; Benthall; Boyle et al.).

---

[3]   In each case microfilm copies of *The Los Angeles Times* archived in the UCLA library were searched for the week after the event and also for several days before and after this period. Articles and advertisements directly related to the earthquake event were photocopied and later analysed for content. However, in this chapter, only significant examples of the style of coverage will be referenced.

America Will Rebuild Japan: Kanto Plain Earthquake 1923

When Tokyo and Yokohama were ravaged by earthquake and fire in 1923, the *LAT* was a more sophisticated media enterprise than it had been in the 1890s. It published Spanish language news summaries to attract the attention of Mexican-origin readers in Los Angeles. It ran its own radio station and advertised its content in a weekly section of the newspaper. The *LAT* celebrated the global potential of the radio technology that was at its command, and it boasted some foreign news correspondents of its own. It seemed to have not only a wider scope of operations around the world but also a more inclusive sense of world citizenship.[4] Perhaps it was for some of these reasons that the *LAT* offered its readers 77,000 words in total (an average of 11,000 words per day) in its first week of coverage of the Kanto Plain earthquake (Table 1).

Nevertheless, only four percent of this news was actually sourced in Japan. Together, Hong Kong, Shanghai, and European capitals supplied more copy than Japan. Overwhelmingly, the editor reported on the quake from locations in the United States (including Hawaii, Guam and Manila). Generally, the reports were produced in-house with scant attention to Japanese news agencies. Even AP supplied less than 29 percent of the copy. This meant that Los Angeles, with 43 percent of copy, was the single most important source point. The *LAT* datelines referred to named correspondents, one (with a Japanese name) located in New York, the others in Los Angeles, Washington, and Kobe. In these ways the *LAT* compensated for the obvious failure of communications from Tokyo and Yokohama. Nevertheless, the *LAT* wrote into its reports its own mastery of communications, 1923-style. It had the latest radio bulletins from British and American shipping and from Shanghai and Kobe. There may have been only one radio operator left in Tokyo, but the *LAT* had his account (*LAT* 3, 4, 8 Sep 1923).

The editor tells this story using the language and themes that Jack Lule found to be characteristic of modern US disaster reporting. This earthquake was a truly horrific disaster and the *LAT* emphasized the terrible devastation wrought by the earthquake using photographs, lists of damaged buildings, reports of the dead, and eyewitness reports. Headlines like "Death toll exceeds 100,000 in Japan's holocaust," "Million homeless and starving in Japan's ruins," and "Flaming river consumes city," emphasized the horror of the catastrophe (3, 4, 8 Sep 1923). A sketch

---

[4]   In 1896, the *LAT* drew attention to the plight of Christians within the Ottoman Empire, while failing to report the Sanriku earthquake. In contrast, in 1923, it reported both the Tokyo earthquake and the inspection of Turkish hospitals by the officials of the Rockefeller Foundation who were considering offering medical scholarships to Turks. *LAT* 2 Sep 1923.

map portrayed the extent of the disaster zone while a cartoon showed the lower slopes of Mt Fuji covered with skulls (3, 5 Sep 1923). Altogether the scope and scale of the disaster amounted to a quarter of the *LAT*'s coverage.

The horrific earthquake conditions were juxtaposed with background stories and images of what ordinary life was like in Tokyo (2, 4, 6 Sep 1923). Science-related stories surfaced, including short bulletins from widespread datelines reporting high tides in Hawaii and seismic readings at US and European observatories (2-4 Sep 1923). Earthquakes, one report declared, were not caused by the weather (5 Sep 1923). A few American buildings in the quake zone survived and this was taken as evidence of superior US building technology (6 Sep 1923), a feat that ignored both the geography of building vulnerability within Tokyo and the long history of Japanese efforts to seek alternatives to European-style masonry buildings, but which may have rested on the putative advantages of steel frame construction (Clancy).

The Japanese government was reported to be active but barely in charge of the scene. Japan's leadership, including the Imperial Family, suffered losses (*LAT* 5, 8 Sep 1923). Cabinet ministers were dead and a new cabinet was sworn in—in the open to avoid falling masonry (4, 6 Sep 1923). There were signs of recovery towards the end of this news week, but many Japanese owed their lives to a fortuitous holiday and not to the planning of the Japanese government (4, 6-8 Sep 1923). 'Korean rebels' were fighting in the streets, so martial law was declared and 15,000 Koreans interned (7, 8 Sep 1923). The *LAT* passed no judgment on this; it was, after all, standard practice to declare martial law and round up trouble makers, as had occurred in San Francisco in 1906. It was only later that the arrest of so many Koreans was found to be unwarranted (8 Oct 1923). This made for dramatic reading, but none of these aspects constituted the dominant story of the Tokyo earthquake; indeed, as the death toll mounted and Japanese battleships sank, these stories were pushed to the corners (6 Sep 1923).

The big story, amounting to 41 percent of copy, was the American relief effort. A cartoon from 4 September shows Uncle Sam offering aid to a stricken Japan (see next page). The cartoon had a competitive note, which is strengthened by the headline in the same edition of the newspaper that "America leads in relief aims." Readers learned that "America will rebuild Japan" as a demonstration of its charity and great civilizing power (4 Sep 1923). The editor hoped that by sending "food ships" instead of "warships" Washington would develop friendly relations with a "humbled Japan" (6 Sep 1923). The relief effort, conducted by the US Navy, was reported as a national matter. President Coolidge declared it was Americans' moral duty to give aid. The US government was reportedly very active, and planned a massive relief effort (2-4, 6-8 Sep 1923).

Hands Across the Sea

Source: *LAT* 4 Sep 1923

But the relief effort was also a matter of local news and transnational competition. Throughout 'Southland,' as the newspaper called Southern California, relief committees sprang into action (3-8 Sep 1923). A broad cross section of Los Angeles society joined the campaign, and the *LAT* reported the responses of the city's clergy and business leaders. The *LAT* collected donations and reported Hollywood fund-raising events. Prominent donors were named. By the end of the week the *LAT*'s relief fund for the Japanese had collected just over 19,000 dollars (4-8 Sep 1923). Los Angeles' Japanese community aimed to raise a half a million dollars (4 Sep 1923). "Nippon girls" volunteered to help the Red Cross appeal, and at a service in Japanese Union Church the city's Japanese community thanked the City of Los Angeles for its efforts during Japan's disaster (7, 8 Sep 1923). This was about competition: as Japanese Americans raised funds in Los Angeles, San Francisco, Chicago, and Honolulu, the question was whether Hong Kong would raise more funds than Los Angeles when each had pledged a half million dollars. The American Red Cross aimed to raise five million dollars (4, 5 Sep 1923).

Californians also learned that their state would have a trade boom in aid to Japan (7 Sep 1923). The entire Californian rice crop had already been sold, along with millions of board feet of timber (5, 6 Sep 1923). An advertisement by a firm of brokers announced that a Portland lumber company had resolved to supply Japan with timber and that its bonds were available to conservative investors (8 Sep 1923). Such news contrasted with the gloom in New York, where the silk exchange was closed and supplies of silk had run out. It seemed that the adverse economic impacts of the disaster in the USA would be confined to the silk industry and to the holders of Japanese bonds, while California enjoyed a trade boom (5, 6 Sep 1923).

Earthquake risk really was shared by Californians. Photographs of Angelinos who were known to be in Tokyo were published. Their fates and the concerns of their relatives were a further substantial component of the news. The *LAT* published lists of Californians known to have been in Japan, announced that ships and individuals were safe or missing, and trumpeted that the daughter of a Hollywood mogul had been pulled from the ruins of her Tokyo hotel (3-5, 8 Sep 1923). After all, it was normal for Californians to be going about their business, proselytizing, or touring in Tokyo.

In the four weeks following 15 September 1923, the *LAT* printed a further ninety-seven articles on the earthquake, comprising over 40,000 words and more than twenty-five images. Reports commented upon the situation in Japan, the extent of the losses, reconstruction efforts, the effects on Japanese industry, and Japan's 'unbroken spirit,' (18, 22, 24 Sep; 1, 5 Oct 1923) but other framings of the post-disaster scene compromised the positive tone of these articles. As typhoons and aftershocks complicated the relief effort, persistent reports declared that Japan's Navy was "crippled," despite the denials of Japanese officials (16, 19, 23, 27, 29 Sep; 3 Oct 1923). By featuring accounts of the disaster from survivors and its own special correspondents, the *LAT* highlighted US quake experiences.[5] While travel writers discussed the exotic practices and superstitions of the Japanese and Chinese, in apparent contrast, architects celebrated Frank Lloyd Wright's modern design of the Imperial Hotel, which had withstood the earthquake (16, 17, 19, 28, 30 Sep 1923). Only one piece offered a scientific explanation for the earthquake, declaring that tsunamis were produced by mass movements in submarine trenches that looked like the Grand Canyon (8 Oct 1923). Stories about communications technology continued to bolster the newspaper's credibility. The *LAT* trumpeted the capabilities of radio for distant communications whilst seizing the opportunity to promote its own radio station and its achievements in delivering the disaster news (30 Sep 1923). Then in a series of "firsts," readers viewed the first pictures of burning Tokyo, the arrival of the first film of the quake in Seattle and San Francisco, its onward journey by warplane and car to Los Angeles and New York and its presentation to Los Angeles audiences, and the publication of an image of the

---

[5] These were intensified as survivors began to disembark from steamers arriving in Pacific coast ports. *LAT* 16-17, 19-20, 23, 30 Sep 1923; 1, 5, 8 Oct 1923.

ruins of the US Embassy in Tokyo, declared to be the first news wire photo sent by radio (17, 18, 21, 25-27 Sep; 3 Oct 1923). Each of these post-disaster framing practices contrasted modern US understandings of earthquakes either with accounts of traditional Japanese practices, or with the Japanese government's failures in delivering modernity, for example, in the form of a tsunami-proof battle fleet.

Nevertheless, the main script continued to be the ongoing fund-raising events in Los Angeles, relief shipments to Japan, and Japanese gratitude for US aid (20, 23-27, 30, Sep; 2, 7 Oct 1923). By the beginning of October, the newspaper's relief fund amounted to little more than 43,000 dollars, but the generosity of the United States was celebrated in an editorial, "America's Way," which declared the USA's hope of rebuilding Japan, a message that was underscored when a Soviet relief ship was turned away by the Japanese authorities amid accusations that it brought propaganda as well as food (21, 22 Sep; 3 Oct 1923). The disaster continued to be reported in terms of opportunities for Californian businesses, and of high hopes for better relations with a humbled Japan.[6]

To summarize, in 1923 the *LAT* told a story of California's intimate relationship with the Tokyo earthquake. It was a local story; of Angelinos caught in the disaster zone, of anxious Japanese residents of Los Angeles reacting to the disaster across the Pacific, of Californians giving aid, and of a booming state export economy. And it was also a national story; the US Navy delivered aid; the US government was reportedly very active; and there were moral lessons for Americans— their charity could make peace with a Japan rendered meek and beholden. Indeed, there were few signs that other nations were involved in the Japanese disaster (4, 5, 7, 8 Sep 1923). The science component of the copy was miniscule. Instead the *LAT* trumpeted its superior communications, which could overcome the disruptions in the quake zone. Together, these individual themes combined to present a mass mediated ritual of the USA's aid program, international networks caught in the disaster, and a foreign government restoring order. In this ritualized coverage, Californians were characterized as generous, caring, Christian, and moral in response to the horrible tragedy afflicting the citizens of Tokyo and Yokohama. In the absence of trans-Pacific jet travel, daily delivery of Japanese newspapers in Los Angeles, satellite news feeds, and internet connections, and with inter-imperial rivalries at the fore, the *LAT* reported the Kanto Plain earthquake as a matter of transnational sympathy, of Californians at risk in the disaster zone, and of a large-scale US relief effort. These issues of the *LAT* from 1923 were already full of dire news concern-

---

[6]   Prospects for US automobile sales in Japan were thought to be low, a Californian firm began to ship houses to Japan, and insurance companies advertised their services to Californians. *LAT* 7 Oct 1923.

ing the state of the world.[7] Between 1923 and 1995, imperial rivalries, war, and the outright occupation of Japan by US forces intervened to rewrite the Japan earthquake narratives in the *LAT*.

An Impossible Script: Depression Era, Wartime, and Cold War Earthquakes

LAT coverage of the Tango quake of 7 March 1927 (Table 1) shows a continuation of the 1923 story lines. AP bulletins report the devastation around Kobe, the rising death toll, and the winter storm that had impeded Japanese relief efforts (8, 10, 11, 13, 14 Mar 1927). President Coolidge offered aid, and encouraged the American Red Cross to send relief to Japan (9, 10 Mar 1927). Americans were again reported as being in the disaster zone, but there were fewer of them. American tourists and a businessman from New York were reported dead, while American missionaries were reported safe (8, 10 Mar 1927). The full range of communications technologies were used to deliver the latest news to Los Angeles and AP again got a beat on the Japanese embassy in Washington (9 Mar 1927). US scientific credibility was asserted with respect to earthquakes. The seismic activity was too much for the needle of one English seismograph (9 Mar 1927). The US would study earthquakes from a new facility in the University of Hawai'i. In announcing this, Captain Paul Whitney of the US Coast and Geodetic Survey also declared that earthquakes are unpredictable and unpreventable, but that when

> any natural peril is understood, protection becomes largely an engineering problem and engineering problems can be solved. Natural phenomena may not be subject to human control but the disastrous effects can be minimized by protective measures based on careful scientific research. (10 Mar 1927)

In these ways, the *LAT* confirms that the USA had a modern understanding of earthquakes, that Americans were in the danger zone, that the newspaper's superior communications network allowed the distant disaster to be reported, and that it was

---

[7] For example, in this week Italy invaded Corfu and 400,000 Fascists seized control of Nuremburg. One cartoon predicted an apocalyptic future for Europe while another advertised a new book by depicting 'Humanity' separated from the Cross by a wall of bayonets, rifles and artillery pieces. The Pan-Pacific Scientific Congress, meeting in Australia heard from geography professor Griffith Taylor that the world had too many people considering its resources. *LAT* 1, 2, 8 Sep 1923.

America's moral duty to be charitable, not only in this instance, but as a matter of national character (13 Mar 1927).

But, for several reasons, the earthquake in 1927 was covered by a fraction of the number of words used to report the Kanto Plain disaster, and both the efforts of Californians to raise relief funds and the rhetoric of the US rebuilding Japan were absent. Partly, the limited coverage was a matter of the smaller scale of this disaster, which resulted in three thousand deaths, a fraction of the total number of fatalities in 1923 (table 1). In addition, the Japanese ambassador in Washington declared that the Japanese government was already active in relief, expressed thanks for the US president's kind offer, but declined to accept it (9, 10, 11 Mar 1927). The Japanese government would not comply with the story line expected of it in the United States. This reluctance on the part of the Japanese government was the result of its humiliating experiences in 1923. Moreover, there was growing US antipathy for Japan following that country's invasion of China. The *LAT* explained to readers that America was entangled in China, along with Japan and Europe, and that peace in China was long delayed. Indeed, readers learned that same week of the march of Japanese forces on Shanghai. Readers also learned that Britain and Japan had agreed to President Coolidge's invitation to negotiate limits to naval buildup and to sign a non-aggression pact.[8] In this context the reporting of the earthquake was a confirmation and continuation of the general thrust of news about Japan: relations were difficult.

For similar reasons, the *LAT* coverage of the Sanriku earthquake of March 1933 amounted to little more than a 1,000 words (Table 1). All of the news came from news agencies and from Japan, and there were few signs of Los Angeles engagement with the disaster. Readers learned that thousands had died but that the victims were being relieved by the Japanese government (3-5 Mar 1933). It was not expected that Americans would give aid to Japan. Readers were informed of whereabouts in China Japan was 'battling for new territory,' and of the fall of the Chinese city of Jehol (3, 4 Mar 1933). They also learned of Japan's troubled financial state—trading was suspended by Japanese banks, and on exchanges in Tokyo, Osaka, and Kobe. The Japanese government sought to know what the US government would do about the financial crisis, but this situation was turned into a moral

---

[8]   This message is illustrated in two cartoons. In "The China egg," *LAT* 11 Mar 1927, a 'Dove of Peace' sits on an egg labelled 'China' atop a nest of swords, and thinks "I wonder if this thing is *ever* going to hatch?" In "An Asiatic laocoon," *LAT* 13 Mar 1927, a dragon labelled 'China' encircles, entraps, and puts the squeeze on figures representing the USA, Europe and Japan. See also *LAT* 7 Mar 1927: "Peril grips Soochow," "Eyes of the world are on Shanghai;" 12 Mar: "Japanese accept bid;" 13 Mar: "New light on China's plight."

lesson for Californians; financial trouble was the upshot of Japan's aggression in China.[9]

The main news stories of the week included clear signs that all of the major nations faced daunting troubles in the Depression. But the swearing-in of President Roosevelt and his immediate actions to stop runs on banks offered hope for turning the corner on the depression in the United States.[10] Readers learned of Filipinos objecting to Japan's war in China, and of their determination to boycott Japanese goods (6 Mar 1933), but this did not signal an end to charity. In the pages of the *LAT* charity was merely redirected. A Navy ball aimed to raise funds for local charities. Despite the forced closure of US bank doors and the new limits imposed on withdrawals, readers were enjoined to give at home. One editorial contended that it would be "unwise" not to give despite the absence of ready cash. Los Angeles residents "can't afford to stand by and see others suffering undeserved misfortune without paying a price in bankrupted humanity far more expensive than the cost of ministering to their needs" (5, 6 Mar 1933). However, Japan had already placed itself outside the reach of American charity.

Coverage of Japanese earthquakes took on new tones during the Second World War. The Tottori earthquake rocked western Japan in early September of 1943 as Australian and US forces scored victories from Burma to Papua New Guinea. In 1943, the newspaper celebrated superior American intelligence and communications in its reports of the war in the Pacific. Thus, in the week of the Tottori earthquake it printed wire photos sent by the US Navy, intercepted Japanese radio transmissions, and the text of a radio broadcast by Adolf Hitler (11 Sep 1943). It is in this fashion that the Japanese quake was reported. Scientists at Caltech and Fordham University recorded heavy tremors near Japan and declared them to have been much stronger than the Long Beach earthquake of 10 March 1933 (11 Sep 1943). Citing a Domei news agency report of 1,400 deaths, an AP bulletin contested a Tokyo Radio broadcast claiming that the earthquake caused only slight damage and minor casualties (12 Sep 1943). In fact, AP is also incorrect: the Tottori quake claimed 12,000 lives (Table 1). In any case, anti-Japanese rhetoric pervaded the news. The newspaper printed some evidence of Japanese atrocities, and advertised the premier of *Behind the Rising Sun*, an RKO Radio movie, with the slogans "They're worse than killers!" and "Know the shocking truth about the

---

[9] A cartoon, "The camp follower," *LAT* 2 Mar 1933, shows a ragged figure labelled "National bankruptcy" pursuing advancing Japanese tanks. See also 6 Mar 1933: "Japanese banks suspend exchange trade."

[10] In the cartoon "There's always a bottom to 'em," Uncle Sam hits "business bed rock" after a long slide and declares "Ah ha! The turning point has been reached at last!" *LAT* 8 Mar 1933.

Japs!"[11] Such rhetoric ruled out sympathy for the earthquake victims. Instead, the newspaper harnessed American generosity to the war bond drive by reporting Los Angeles clergy, a New York striptease artist, and 50,000 onlookers at a military parade in the city each doing their bit to raise funds (11, 12, 14, 15 Sep 1943).

The Tonankai earthquake of 7 December 1944 struck as US bombers flew missions over Tokyo. This week was an anniversary of the attack on Pearl Harbor, and, as the *LAT* took stock of the accumulating damage to Japan, it was clear that the tide had turned (11, 13, 14 Dec 1944). Under the title "Imperial Tokyo, Heart of Japan and No. 1 Pacific Target," readers found a full-page aerial photograph of Tokyo emblazoned by the *LAT* with US bombers and bombs, and the message that "as a target, Tokyo is among the best" (13 Dec 1944). Advertisements appeared for Spencer Tracy's new film *Thirty Seconds Over Tokyo* (8 Dec 1944). While one Angelino stated that he was keen to visit Tokyo with his bomber, others dropped their empty beer bottles over the city. Indeed, as bombs fell near the Imperial palace, the earthquake in central Japan seemed merely a sideshow to the destruction being wrought on the other side of the Pacific Ocean (7, 12, 14 Dec 1944).

Again, there was apparent controversy over the extent of the disaster. While reports from Tokyo made little of the earthquake, experts in London called it a quake of "catastrophic force" and Dr. Beno Gutenberg of Caltech, shown with his seismographic record of the quake, contended that the quake was accompanied by tidal waves and was "as great as the devastating 1923 Tokyo earthquake": an engineer at the Department of Water and Power, San Fernando Valley found the signature of the tremors on the record of the city's water reservoir level (8 Dec 1944). Subsequent Japanese news agency reports acknowledged damage to industry in the center of Honshu, but claimed that damage was minimized by precautions already taken against US bombing (9 Dec 1944). From Colgate University, geologist Dr. Harold Whitnall attributed the earthquake to B-29 bombings of Tokyo, arguing that giant bombs dropped in or near volcanoes should produce eruptions, earthquakes and tidal waves. But a seismologist at Fordham University offered a hot denial: "The effect of a bomb compared with the effect of an earthquake is as a flea is to an elephant" (9 Dec 1944). Still, despite the sophisticated remote sensing technologies available to the Allied scientists it proved hard to determine the extent of the damage wrought by the earthquake amid the wreckage caused by the bombs

---

[11]   One advertisement proclaims that "To the vicious Japs, a woman is less than nothing!—a woman is only a female—to be bartered, tortured, or tossed aside—be she wife, daughter or just common Geisha girl!... That's "chivalry" among the "Sons of Heaven", the vilest villains the world has ever known!" *LAT* 10 Sep 1943: "Japs slew own patients on Attu, diary discloses;" 15 Sep 1943: "Film drama dissects nature of Japs."

raining down on Japan. Curiously, there was no report of the Mikawa earthquake of 12 January 1945 in the *LAT*, probably indicating its failure to register as a significant wartime event, rather than a failure of detection.

Meanwhile, in Washington DC, Congress discussed the prospect of returning interned Japanese to their homes. In fact, eight hundred Japanese-Americans had been freed, including Mr. Fukuda and his family of Orange County (11 Dec 1944). The neighboring ranchers gave the family a warm welcome, and Fukuda's daughters practiced for their part in an Anaheim Lutheran Church Christmas performance. But California representatives objected to the release of Japanese-Americans during the war on the grounds that "it would cause riots, turmoil, bloodshed, and endanger the war effort" (12 Dec 1944). As Christmas approached, and the Salvation Army announced its lavish plans for giving in Los Angeles, the bounds to Californian generosity to the Japanese in Japan, South East Asia, and in the USA were a matter of intense debate, a debate that continued into January 1945, when the Mikawa earthquake struck Japan (9 Dec 1944; 13, 15 Jan 1945).

Two further earthquakes devastated parts of Japan in the immediate postwar era, while US military forces occupied Japan. Thanks to the occupation, the devastation of the Nankaido event of 20 December 1946 could be spelled out, and the *LAT* used AP and United Press International (UPI) wire photos and bulletins to express the sensational aspects of the disaster for the Japanese victims (21, 23, 26 Dec 1946). The disaster area was mapped and reports noted the sequence of, firstly, the earthquake, which caused destruction along a 240 kilometer belt of central Japan, and then, within fifteen minutes, a two meter high tsunami, which was funneled through the Inland Sea, inundating the coast, including Osaka (21, 22, 25 Dec 1946). Among the occupying forces, only one British casualty was reported, but thousands of Japanese residents were affected and up to 15,000 houses were swept away. Japanese and Allied relief teams were reported to be active. The news bulletins regularly made the Kanto Plain earthquake of 1923 the benchmark for the Nankaido disaster. Nankaido was "the most disastrous earthquake and tidal wave in 23 years;" the newspaper ran a background piece on the 1923 disaster, and reported that the Japanese would ask for US aid, as they had in 1923 (21, 22 Dec 1946). However, one UPI bulletin drew a direct parallel with the US bombing raids, noting that the tsunami arrived on the same path as the one taken by bombers headed for Tokyo in a series of attacks on that city in 1945 (20 Dec 1946). Again, Caltech seismologists were credited with recording the tremors, including less consequential earthquakes in Scotland, El Salvador, and Nicaragua, indicating the global capabilities of earthquake detection services, themselves relevant to the monitoring of atom bomb tests during the Cold War (21, 22, 26 Dec 1946).

In December 1946, Los Angeles' second postwar Christmas, there were calls for world peace and for charity. The *LAT* reported numerous Christmas appeals and hospital visits. The editor highlighted the Salvation Army's Christmas dinners and called upon readers to "make it a feast, not just pot luck" (22, 24-26 Dec 1946). Christmas giving was celebrated as an American act, "charity rackets" were decried, and a cartoon entitled "No exchanges, please," showed a gift, suitably wrapped for the season and labeled "peace," resting on a globe. Both the Pope and President Truman prayed for peace on Christmas Day (23-25 Dec 1946). As tensions rose with the Soviet Union, the *LAT* wrestled with the troubling question of whether there could, indeed, be a lasting peace. There was a Christmas message from Tokyo, and one Los Angeles church gave relief for Japan, but, even with the ideal of America rebuilding Japan back on the US government's agenda, and so soon after atom bombs leveled Japanese cities, it was left to the US Army to deliver aid to the earthquake victims. Indeed, the *LAT* reports followed the delivery of aid to the American soldiers who were in the quake zone.[12] In the pages of the *LAT*, American charity was still best directed to needy citizens at home; the US government would provide for Americans in Japan.

In late June, 1948, as US "raisin bombers" flew to defeat the Soviet Union's Berlin blockade, an earthquake in the west of Japan took 3,800 lives (28 Jun 1948). The *LAT* published bulletins describing the extent of the disaster but aid to Japan proved difficult to justify to readers (29, 30 Jun; 1, 2 Jul 1948). This was certainly not a matter of charity no longer being celebrated. As the American bombers relieved Berlin, the Californian Red Cross met to re-elect its president and to present awards (28 Jun; 1 Jul 1948). With American relief needed to fight communism, and aid to China cut, the question of aid to Japan remained controversial. In China, students protested against US aid to Japan. At the same time, accusations of wartime torture were made against a Japanese man on trial in the USA, and the Vatican denied that Emperor Hirohito planned to convert (28 Jun; 2, 4 Jul 1948). It seemed that Japan was not going to be rebuilt by America. In mid-1948, as the Cold War intensified, the Japanese were still considered to be outside the community of deserving victims.

The Kobe Earthquake 1995

Southern Californians could have read 53,000 words on the Kobe quake in the week after the event (Table 1). In 1995, the *LAT* boasted 'global coverage'—its

---

[12] Louis Spaeth offers a Christmas message from Tokyo, in which he declares that "One gets much satisfaction from helping here in the rebirth of a nation and in guiding a people toward a true desire for freedom and the will to live in peace with the world." *LAT* 22, 23, 26, 27 Dec 1946.

team of reporters and correspondents covered the globe, and agency sources were now a thing of the past. Web-based news sources were available to readers, though these turned out to be AP and Reuters feeds (18 Jan 1995). In fact Japanese datelines were more important than they had been in 1923, but US datelines continued to dominate. Curiously, Melbourne and San Diego featured among the datelines: Japanese sportsmen and women at the Australian Open and at the America's Cup yacht racing were distraught because their efforts to call home met with no success—the lines were down in Japan (18, 22 Jan 1995). The fact that emergency public telephone services were available to the Japanese authorities was a matter of little help to those anxiously trying to contact relatives. The *LAT* signaled that Japanese residents of Los Angeles still felt very connected to their homeland. They were raising funds and were deeply concerned, and were shown reading Japanese language newspapers featuring news of the disaster (18 Jan 1995). So, the old story line of the *LAT*'s superior communications network was told again, this time in the context of new ideas about globalization and the expectations of personalized global communication systems.

About half of the coverage related directly to the effects of the quake in Japan. Themes of total devastation, relief, rescue, and rebuilding featured prominently, accounting for 39 percent of coverage. A new feature of this reporting was that *LAT* correspondents offered eyewitness accounts of the devastation, relief, and rescue work. Images of twisted buildings, such as a Mitsubishi Bank office, and broken bridges and roads conveyed the alarming extremity of the convulsions from a street-level perspective. Journalists personalized the disaster by reporting the hopes and fears of Japanese survivors in a series of articles describing returns and rescues: a woman was shocked to tears when she saw her ruined home; a family erected a shrine on the site of the house to remember loved ones (19, 21 Jan 1995). In contrast, the Japanese government was criticized by the opposition parties for its slow relief response. The *LAT* juxtaposed claims that the mobilization of soldiers for relief work was slow with a story about queues of well-heeled residents forming outside the precinct of a major gang leader who was dispensing food and water (19, 22 Jan 1995). In these ways, Japanese in the disaster zone were inscribed as citizens who warranted both American sympathy and better public service from the Japanese government.

Once again there were Angelinos in the disaster zone as victims, survivors, and relief workers, and readers were informed of how to help. Their aid could be donated to the Salvation Army or World Vision, but also to the American Red Cross or Operation USA, or to the Japanese Village Plaza or Japanese-America Society of Southern California (18, 19 Jan 1995). One letter to the editor justified aid from Los Angeles in terms of reciprocity.[13] Angelinos made donations and their hearts went out to Kobe but it was in fact difficult to send aid across the Pacific Ocean (21, 23 Jan 1995). Japanese authorities would not accept food or clothing. A medical team, boasting Northridge experience, got to Kobe but was only permitted to perform surgery after much lobbying. The parents of one Los Angeles woman, devout Baptists, struggled to get her body home from Japan (18, 19, 21, 23 Jan 1995): the US ambassador stepped in to cut through uncaring Japanese bureaucracy and to prevent the cremation of her remains. Thus, despite the admirable character of Angelinos, new kinds of barriers emerged in what had become more of a two-way aid relationship. Los Angeles residents shared the Kobe quake but in a fraught fashion, as frustrated donors, and experts whose offers of help were turned down.

Indeed, in 1995 Southern California had a particular relationship with the risk of earthquake, since Los Angeles was commemorating the first anniversary of the Northridge earthquake as the news of the much greater disaster in Kobe began to spread. The *LAT* published memories of Northridge survivors, covered President Clinton's visit to Los Angeles to mark the anniversary, and described the state of recovery in Northridge a year on.[14] Los Angeles psychologists had been offering therapy to those affected by the Northridge quake and they reported a new wave of fears following news from Kobe. They headed to Japan to advise Japanese psychologists. Los Angeles geologists and politicians warned that a quake of the same magnitude as the Kobe one in Los Angeles would be much more devastating than the Northridge event had been. Nevertheless, Mayor Riordan erected a billboard with the slogan "You can shake LA but you can't break it," a strident message of earthquake denial out of step with the expert commentary in the *LAT* (18 Jan 1995).

There was, in fact, considerable expert commentary on the lessons of the Kobe and Northridge disasters. Addressing Northridge, the *LAT* did its best to encourage

---

[13] She writes: "In 1994 the Japanese extended a hand in our time of need. Japanese companies donated 3 million dollars to victims and relief efforts after the Northridge quake.... Helping earthquake victims is an international job." *LAT* 20 Jan 1995: "Returning the favor."

[14] See for example *LAT* 17 Jan 1995: "Clinton visits LA: A caring government needed," "Memories of the quake," "Quake-safe buildings," "Street of shaken dreams," "Seismic future: Shaky future or a false alarm?" "We learned planning is no luxury," "We're petrified that when our insurance runs out we'll lose the house," "Poll: Many say lives are pretty much back to normal;" 18 Jan 1995: "Survivors somberly look back on year since quake," "Quake's toll in Japan shakes LA;" 20 Jan 1995: "A crystal bowl's legacy after Nazis and Northridge quake," "Dollhouses stomped into the ground;" 21 Jan 1995: "Riordan to lead lobbying group seeking US funds."

its readers to prepare for a bigger quake in the future (the "big one"). The limitations of earthquake insurance were discussed, and it was abundantly clear that California was not coping well with its own year-old disaster. The engineering community was reported as being sharply divided over the extent of the threat to steel frame buildings that had been revealed by the Northridge quake, and owners of buildings were still waiting for new standards in steel construction to be set and enforced by the City of Los Angeles. Similarly, damaged schools in Los Angeles were still awaiting the promised state funding to carry out repairs, and there was controversy over whether a damaged hospital should have been allowed to remain open (17, 19, 20, 23 Jan 1995). Finally, in a blow to the Keynesian ideals of government security, ten suspects were indicted for allegedly defrauding the US government of millions of dollars following the 1992 riots, the 1993 Malibu fires, and the 1994 Northridge earthquake in a major case of disaster loan fraud (17 Jan 1995). Californians were invited to question their smug assumptions about Los Angeles' earthquake preparedness and mitigation work (19, 21 Jan 1995).

The *LAT* wrote of the Kobe earthquake as a Pacific Rim disaster which could be understood using corporate and technical expertise. US sport psychologists were needed in Melbourne and San Diego. US geologists and engineers were in Japan and they were appalled by Japan's engineering "mistakes" and lack of preparedness. US scientists declared the quake to have been smaller than first reported by Japanese scientists and presented the latest ideas about the causes of the earthquake (18, 19 Jan 1995). Addressing the Kobe quake, the editor asked for a new level of cooperation between the US and Japan: Japanese authorities would need to reconsider engineering standards in cities built on made ground, highway bridge designs, and relief preparations (18, 20, 22 Jan 1995). Security needed to be redefined in a global context, said Washington-based Worldwatch Institute (23 Jan 1995).

With the costs of rebuilding estimated at 50 billion US dollars it was clear that the Kobe earthquake was an economic catastrophe for Japan. The earthquake restricted trade through Japan's important port city for months, keeping factories and their suppliers across a broad region of West Japan closed, and disrupting global shipping. This 'economic fallout' was understood in terms of points of difference in law and institutions. The US government had been pushing for market liberalization around the globe, and the *LAT* noted Washington's earlier calls for Japan to open construction contracts to foreign bidders (18, 20, 23 Jan 1995). Non-tariff barriers limited the involvement of US firms in rebuilding work. The quake was a matter of US-Japan relations, and here the US had the moral high ground: its experts would have delivered buildings and roads which could have withstood the earthquake better. A technical report compared bridge construction specifications in California and Japan. Thus, with the rebuilding of Kobe likely to be a drawn out and expensive process, the *LAT* highlighted the issue of whether foreign firms would have access to the construction work (19 Jan 1995). Moreover, the disaster would send Japanese firms overseas to look for more secure locations for manufacturing plants, a move that was sure to benefit the US economy. Readers learned that while Japan's socialization of risk, despite its merits, was unacceptable in the United States, Japan's caps on insurance liabilities would insulate insurance companies and prevent a global insurance meltdown. Instead, the Nikkei dropped hard (18 Jan 1995). Wall Street—once "awed by Japan"—now had nothing to fear and the Dow Jones closed higher.

Conclusion

This chapter analyzed reports of nine Japanese earthquakes in one newspaper carefully chosen for its longevity and regional readership within the USA. It confirms well-known aspects of the modern earthquake news story: narratives of science and communications technology helped to authenticate and lend authority to the newspaper's account of the disaster; a myth-like story line emphasized the completeness of destruction, the awesome power of a capricious God or Nature, the fortitude of survivors, hope reborn, and moral relationships between humans; and reports featured public performances of identity and relationship in the form of relief programs. Significantly, the chapter analyses the geographical imaginations used by the *LAT*, in terms of a moral tale of relationships going beyond US borders. Southern Californians were constructed in these stories as experts, aid givers, suffering relatives—in short, as moral citizens of the world.

Analysis reveals continuities in coverage between the Tokyo 1923 and Kobe 1995 earthquakes. In each case the *LAT* described Japan as a bizarre combination of cutting-edge modernity and centuries-old tradition (17, 21 Jan 1995). It counted the scale of the disaster in terms of a human death toll and the destruction of landscapes. Images of destruction and death continued to be fitted around prepaid advertisements for men's suits and women's underwear. The *LAT* related its readers to the earthquake using stories about communication technology, scientific experts, teams of fund-raisers and rescuers, politicians, state authorities, victims, and survivors. These stories were constituted in terms of drama, ritual, and morality. Readers were told about where the story was compiled, under which conditions, and how authoritative it was: they could rely on the *LAT*. The mobilization of Angelinos in a dramatic media ritual of raising and sending relief has been a central pivot of the *LAT* disaster story. The lineaments of this story—the fact that Angelinos of all kinds participate, its Christian moral code, the gratitude of the US and recipient governments, the sympathy that goes with the generous giving, the sure knowledge that there but for the grace of God goes Los Angeles—remained intact. Californians were legitimately at work in Japan, Japanese Americans were caring and respectable US citizens, and Californians led the world as charitable aid donors.

However, the newspaper's representations of Japanese disasters were mediated by the editors' sense of the USA's geo-economic and geopolitical relations with Japan. The history of US-Japan relations has not always been supportive of US aid to Japan. At first, Japanese earthquakes were rendered using news agency bulletins as devastating tragedies for a distant and foreign land. In 1923, the Tokyo disaster was interpreted as an opportunity to build better relations with Japan in the context of imperial rivalries, and the *LAT* reported the US relief campaign for Japan in detail. But by 1927, the framing of the earthquake as a matter of America's generous relief campaigns and capacity to rebuild Japan was no longer plausible since the Japanese government had refused US aid, considering it to be unnecessary and patronizing interference. Increasingly, Japan was perceived in the USA as unworthy of aid. Thus, coverage of Japanese earthquakes was much reduced and largely comprised news agency bulletins, while Californian generosity was redirected to those closer to home. Instead, Japanese earthquakes were presented almost as collateral to warfare. The remote sensing of Japan using seismographic equipment, radio transmission interception, and aerial photography, became an important theme in the newspaper coverage. As the Second World War progressed it became difficult to establish whether damage was the result of tectonic processes or of bombing raids. The citizenship of the Japanese residents of Los Angeles was reworked. A positive role was available to them, but only when this coincided with the dominant framing of US citizenship and identity relative to Japan.

There were signs of an intensifying globalism in reports of the Kobe earthquake of 1995. The *LAT* was by then more capable of writing up this international disaster in-house. There was a return to a full relief campaign, but one now framed within the ideals of neoliberal globalization. The economic effects were then interpreted in terms of foreign direct investment, rather than merely according to trade. Reports featured a clash of national expert systems that was missing in the 1923 reports. Personal accounts of tragedy, survival, grief, sympathy, and charity individualized experience of the earthquake. Both the editor and Angelino readers seemed to have developed heightened expectations of involvement with the Kobe quake, but, curiously, these were frustrated. Access to Kobe was restricted, adversely affecting US-Japan relations, and justifying US government policies. Simultaneously, in the *LAT*'s pages, Washington's role was reduced. Thus globalization, as a neoliberal project, was written into the narrative as the US policy agenda in Japan, replacing the US' (counter) imperial project of 1923.

In these ways, the stories analyzed here reflect changing perceptions of the role of the US government in the world and can be seen as helping to constitute, name, and discuss that changing role for Southern Californians. Whether these recent changes reflect a changing world (the editors are aware of a new readership and new patterns of human activity) or are attempts to constitute a new world (the editors seek to constitute their readers in these terms in order to globalize their newspaper and retain market share) can only be guessed at using the data sources and methodology employed in this chapter. Both probably hold true to some extent; the editors and globalization processes are probably co-constitutive within the *LAT*'s pages, but limits to this joint process can be quickly revealed.

Curiously, the government of Japan was not constituted as the effective actor in charge of any of these earthquake situations. The Japanese government's role was always shadowy: it was either poorly prepared, unable to act, or obstructionist. Local contingency plans and preparations were only discussed in the context of their failure. So, the *LAT* opened an action space for Southern Californians in distant disaster zones by invoking a silence on, or at least a writing-down of the capabilities of the Japanese government. In contrast, Angelinos were portrayed as key actors in the drama, and in 1995 they were joined by Japanese survivors struggling to cope with the tragedy. Angelinos, sometimes inclusively framed as Japanese-Americans, were similarly made subjects and objects of globalization in these narratives. Both the federal government in Washington DC, and, increasingly, Southern Californians, were in center frame.

It could be argued, then, that the general disaster narrative as identified in the *LAT* seems designed to avoid any attention to the social costs of disaster: charity is enough; socialization of the costs of earthquakes is not required in the USA, where everyone pays their own way; martial law and repressive acts by governments are legitimate in such emergency circumstances. Subsequent reports may have featured analysis of disaster preparedness and the allocation of the burdens and costs of the disaster, but in the week immediately following each event the *LAT*'s reporting effectively blocked any rethinking of local arrangements, any inquiry into response or preparation frameworks. The *LAT* created value from earthquake disasters, but this accrued in specific forms: in terms of readership, advertising revenues, and profile for the newspaper; in terms of opportunities to bolster current geographical imaginaries and roles for readers. It did not accrue, at least in the immediate aftermath of disaster, in terms of a critical journalism focused on analyzing preparedness, relief efforts, the socialization of costs, or the sustainability of communities. A disaster is always a key political threat, but in the event of a local earthquake disaster, Californians' reactions would be preconditioned (Barry; Barry et al.). By 1995, this press seems to have narrowed the human response to earthquakes: charity is enough and may eventually become superfluous; governments and experts can be relied upon.

As environmental historians research cultures of disaster in comparative perspective (Mauch et al.), the changing lineaments of a US culture of foreign disasters needs to be brought into the spotlight. This chapter finds that, in the context of earthquake disaster news stories about Japan through to 1995, despite geographic learning, intensified interactions with distant places, and the enhanced news gathering capabilities of the newspaper, the *LAT* consistently built its coverage around the issue of whether Japan was worthy of US aid. It enjoined Angelinos to give to those who were worthy recipients of aid, and worked to redirect US charity to others when the Japanese were thought to be unworthy. This highlights the significance of geographical imaginaries within media framing in shaping expected responses to distant disasters.

Works Cited

Anderson, Benedict. *Imagined Communities: Reflections on the Origin and Spread of Nationalism.* London: Verso, 1983.

Appadurai, Arjun. *Modernity at Large: Cultural Dimensions of Globalization.* Minneapolis: University of Minnesota Press, 1996.

Bankoff, Greg. "Rendering the World Unsafe: 'Vulnerability' as Western Discourse." *Disasters* 25.1 (2001): 19-35.

Barry, Andrew. *Political Machines: Governing and Technological Society.* London: Athlone Press, 2001.

Barry, Andrew and Don Slater. *The Technological Economy.* London: Routledge, 2005.

Benthall, Jonathan. *Disasters, Relief and the Media.* London and New York: Tauris, 1993.

Bird, S. Elizabeth and Robert W. Dardenne. "Myth, Chronicle and Story: Exploring the Narrative Qualities of News." *Social Meanings of News: A Text-Reader.* Ed. Dan Berkowitz. Thousand Oaks: Sage, 1997. 333-49.

Birkland, Thomas A. *Lessons of Disaster: Policy Change After Catastrophic Events.* Washington, DC: Georgetown University Press, 2007.

Bolin, Robert and Lois Stanford. "Constructing Vulnerability in the First World: The Northridge Earthquake in Southern California, 1994." *The Angry Earth: Disaster in Anthropological Perspective.* Eds. Anthony Oliver-Smith and Susanna M. Hoffman. New York: Routledge, 1999. 89-112.

Boyle, Elizabeth Heger and Andrea Hoeschen. "Theorizing the Form of Media Coverage over Time." *Sociological Quarterly* 42.4 (2001): 511-27.

Clancy, Gregory. *Earthquake Nation: The Cultural politics of Japanese Seismicity, 1868-1930.* Berkeley: University of California Press, 2006.

Cosgrove, Denis and Veronica Della Dora. "Mapping Global War: Los Angeles, the Pacific and Charles Owens's Pictorial Cartography." *Annals of the Association of American Geographers* 95.2 (2005): 373-90.

Cottle, Simon. "Mediatised Rituals: Beyond Manufacturing Consent." *Media, Culture, and Society* 28.3 (2006): 411-32.

Couldry, Nick. *Media rituals: A critical approach*. London: Routledge, 2003.

Davis, Mike. *City of Quartz: Excavating the Future in Los Angeles*. New York: Random House, 1992.

Dayan, Daniel and Elihu Katz. *Media Events: The Live Broadcasting of History*. Cambridge, MA: Harvard University Press, 1992.

Hallin, Daniel C. *We Keep America on Top of the World: Television Journalism and the Public Sphere*. London: Routledge, 1994.

Hannah, Matthew. "Torture and the Ticking Bomb: The War on Terrorism as a Geographical Imagination of Power/knowledge." *Annals of the Association of American Geographers* 96.3 (2006): 622-40.

Hewitt, Kenneth. "The Idea of Calamity in a Technocratic Age." *Interpretations of Calamity: From the Viewpoint of Human Ecology*. Ed. Kenneth Hewitt. Boston: Allen and Unwin, 1983. 3-32.

Heyer, Paul. *Titanic Legacy: Disaster as Media Event and Myth*. Westport Connecticut: Praeger, 1995.

Hutchinson, John F. *Champions of Charity: War and the Rise of the Red Cross*. Boulder, CO: Westview Press, 1992.

Lule, Jack. *Daily News, Eternal Stories: The Mythological Role of Journalism*. New York: Guilford Press, 2001.

Lynch, Jake and Annabel McGoldrick. "Peace Journalism: A Global Dialog for Democracy and Democratic Media." *Democratizing Global Media: One World, Many Struggles*. Ed. Robert A. Hackett and Yuezhi Zhao. Lanham, MD: Rowman and Littlefield, 2005. 269-312.

Mauch, Christof and Christian Pfister. *Natural Disasters, Cultural Responses: Case Studies Toward a Global Environmental History*. Lanham, MD: Lexington Books, 2009.

Moeller, Susan D. *Compassion Fatigue: How the Media Sell Disease, Famine, War and Death*. New York: Routledge, 1999.

Oliver-Smith, Anthony. "Peru's Five-hundred-year Earthquake: Vulnerability in Historical Context." *The Angry Earth: Disaster in Anthropological Perspective*. Eds. Anthony Oliver-Smith and Susanna M. Hoffman. New York: Routledge, 1999. 74-88.

Oliver-Smith, Anthony, and Susanna Hoffman (eds.). *The Angry Earth: Disaster in Anthropological Perspective*. New York: Routledge, 1999.

Quarantelli, E.L. "Disaster Crisis Management: A Summary of Research Findings." *Journal of Management Studies* 25.4 (1988): 373-85.

———. "Ten Criteria for Evaluating the Management of Community Disasters." *Disasters* 21.1 (1997): 39-56.

Rosenblum, Mort. *Coups and Earthquakes: Reporting the World for America*. New York: Harper and Row, 1979.

Rotberg, Robert and Thomas G. Weiss (eds.). *From Massacres to Genocide: The Media, Humanitarian Crises, and Policy-making.* Cambridge, MA: World Peace Foundation, 1995.

Scott, Allen J. and Edward. W. Soja (eds.). *The City: Los Angeles and Urban Theory at the End of the Twentieth Century.* Berkeley: University of California Press, 1996.

Singer, Jane B. "More Than Ink-stained Wretches: The Re-socialization of Print Journalists in Converged Newsrooms." *Journalism and Mass Communication Quarterly* 81.4 (2004): 838-56.

Soja, Edward W. *Thirdspace: Journeys to Los Angeles and Other Real-and-Imagined Places.* Cambridge, MA: Blackwell, 1996.

Steinberg, Theodore. *Acts of God: The Unnatural History of Natural Disaster in America.* Oxford: Oxford University Press, 2000.

Whitney, Charles D. and Lee Becker. "Keeping the Gates for Gatekeepers: The Effects of Wire News." *Journalism Quarterly* 59.1 (1982): 60-65.

Wilke, Jürgen. "Historical Perspectives on Media Events: A Comparison of the Lisbon Earthquake in 1755 and the Tsunami Catastrophe in 2004." *Media events in a global Age.* Eds. Nick Couldry, Andreas Hepp, and Friedrich Krotz. London: Routledge, 2010. 45-60.

Winder, Gordon M. "Mediating Foreign Disasters: *The Los Angeles Times* and International Relief, 1891-1914." *Historical Disasters in Context: Science, Religion, and Politics.* Ed. Andrea Janku, Gerrit Schenk, and Franz Mauelshagen. Oxford: Routledge, forthcoming.

Forgetting the Unforgettable: Losing Resilience in New Orleans[1]

Craig E. Colten

With the passing of Hurricane Katrina's five-year anniversary in August 2010, New Orleans hosted numerous commemorative events. A central theme at these gatherings has been the community's resilience, its ability to bounce back and re-build after an unquestionably devastating event. From this five-year vantage point, it is possible to look back and measure the progress of the city in restoring its infra-structure, its population, and its economy. The Brookings Institution has offered a relatively optimistic view of the city's recovery, proclaiming that the number of blighted and abandoned properties is declining rapidly, the population has climbed back to approximately 78 percent of the pre-storm level, and the economy has di-versified. It also explicitly reports that greater New Orleans has become more resil-ient, largely through "an unprecedented rise in civic engagement" and reform in civic institutions (Liu and Plyer 6). While these assessments are accurate, another way to gauge the city's resilience is to compare its recovery after Katrina with its recovery after Hurricane Betsy in 1965. The memory of Katrina is still prominent and continues to drive recovery efforts. But, will that memory persist until the next storm? The longer-term perspective afforded by the forty-year interval between Betsy and Katrina allows us to examine the erosion of social memory and conse-quent loss of resilience.

When Betsy made landfall in September 1965, New Orleans had much less structural protection and the storm drove more powerful winds through the city than its notorious younger sister. Yet despite the massive investment in hurricane protection levees since 1965, Katrina overwhelmed these barriers to unleash a havoc as yet unknown to a city all too familiar with tropical cyclones. This chapter examines changes in three areas of community resilience, making the case that New Orleans lost resilience between 1965 and 2005. Although the Brookings Insti-tution asserts an increase in resilience since Katrina, my contention is that in the years leading up to the more recent storm, the city's capability to contend with massive disruption had diminished, and this fact contributed to the calamity in

[1] Acknowledgments: I would like to thank the Community and Regional Resilience Institute for funding much of this research, and the capable research assistance from Amy Sumpter and Al-exandra Giancarlo.

2005. This longer-term view begs the question: Will the new resilience assembled during the five-year interval since the storm survive the gradual waning of public interest? Will this newfound resilience endure?

Comparing Two Storms

If we compare some of the most basic elements of Betsy and Katrina with their respective impacts, the results are disorienting. Betsy delivered wind speeds within the city of 125 miles per hour, at least before power failed and disabled monitoring devices (National Hurricane Center 1965b). Approximately 627,000 citizens lived within the city limits at the time. Modest five-foot tidal levees protected residents in the eastern sections of the city, and more substantial levees armored the Lake Pontchartrain shore and areas along the Industrial Canal, standing 9.5 feet and 7 feet high respectively. Hurricane monitoring capabilities included radar and hurricane hunter aircraft, but forecast accuracy meant that the abilities of weather bureau personnel were limited to the issue of a geographically vague hurricane watch that stretched from near Biloxi, Mississippi to Matagorda Bay, Texas, twenty-four hours before landfall (National Hurricane Center 1965a).

By contrast, in 2005, the National Hurricane Center's prediction seventy-two hours before landfall was incredibly accurate. As Katrina moved on shore, the highest sustained wind speeds measured in New Orleans were only 87 miles per hour; higher velocity gusts prompted the Hurricane Center to conclude the storm's intensity was no greater than a category two level (96-110 mph; National Hurricane Center 2005). Since 1965, population had fallen to 484,000, according to the 2000 census, so there were fewer people in the city at risk. Between the two storms, the Corps of Engineers had led a massive levee building project that added levees, standing over fourteen feet high, around the most flood-prone eastern and northern edges of the city (Colten 2009).

Both storms unleashed serious consequences, and the pain and suffering of those who endured these events was deeply felt and longlasting. Nonetheless, when considering the impact at a community level, Betsy's devastation, while transformative, was less costly by most measures. Betsy flooded only 43 percent of the incompletely armored city, compared to 80 percent of the city inundated by Katrina, despite the massive levee system that had been built in the interim. Approximately 15,000 homes suffered major damage in 1965, which pales in comparison to the 105,000 homes seriously damaged in 2005. And the most painful measures were the 81 fatalities in 1965, compared to some 1,300 in 2005 (USACE 1965; National Hurricane Center 2005).

The restoration of basic infrastructure and civic functions offers a measure of resilience during the short-term recovery period. Following Betsy, the local newspaper, with typical booster enthusiasm, reported that "all big stores" would open two days after the September 9 landfall (*NOTP* 9 Oct 1965, 1). By September 12, air, rail, and bus services were at near normal levels and most public transit services had resumed operation—with the exception of several still-flooded neighborhoods (*NOTP* 12 Sep 1965, 1, 9). By September 15, the *Times-Picayune* reported that conventions had resumed, while 75 percent of the population had electrical power, and two thirds had had their phone service restored (15 Sep 1965, 8). A week after the storm, four public and thirty Catholic schools were offering classes and the city rescinded the boil-water order (16 Sep 1965, 24, 29). Ten days after landfall, only small areas still had standing water (9 Sep 1965, 13); and by the twentieth of the month, 90 percent of the city had electrical power (20 Sep 1965, 6). Restoration of power and potable water supplies were factors that enabled people to return to their homes, and by September 22, only 5,300 people remained in shelters. The last of the city schools finally reopened on October 5, less than one month after the storm (5 Oct 1965, 1). The public service utility company took out a large advertisement proclaiming that "New Orleans is Back in Business" on the one-month anniversary of the storm (9 Oct 1965, 2).

Following Katrina's August 29 landfall (and the secondary flooding caused by Hurricane Rita a few weeks later), the restoration of services was much more protracted. Closing the levee breaches and pumping the city dry took six weeks. City and private sector crews restored the following services to most neighborhoods after lengthy efforts (to the 90 percent level): potable water on December 16, 2005; sewerage service on May 5, 2006; and electricity on June 21, 2006 (New Orleans). Four months after the storm, the population approached 37 percent of the pre-storm total, and it had rebounded to about 48 percent by October 2006 (LDHH). Most schools (private and public) reopened for the spring term in January 2006—although the University of New Orleans started holding classes as early as October. The hospitality industry resumed hosting major conferences after a nine-month moratorium. At the one-year mark, public transit and hospital services remained at less than half capacity compared to pre-Katrina levels (Liu and Plyer). The Corps of Engineers had targeted the start of hurricane season, 1 June 2006, as the date for completing the initial levee repairs and flood gates on the outfall canals. Although they did not meet that deadline, the repairs were in place before the end of the 2006 hurricane season (USACE 2006a, 2006b). Given the scale of disruption and the length of the emergency period, recovery was more or less on track, but it was taking longer due to the extensive initial damages (see Kates et al.; Colten et al. 2008). The fact that a lesser storm, one which encountered more substantial protection structures, caused such extensive damages suggests a serious loss of resilience between 1965 and 2005.

Defining Resilience and the Role of Social Memory

After calamitous events, leaders commonly call on their constituents to learn their lessons and to rebuild their communities in a better and safer way. This was certainly the case after Hurricane Betsy. Louisiana's governor, John McKeithen, led the chorus. Testifying before congress two weeks after the storm, he proclaimed the desire "to establish procedures that will someday in the near future make a repeat of this disaster impossible" (McKeithen 32). While not using the term resilience in the 1960s, McKeithen was alluding to one element of making coastal Louisiana more resilient—namely reducing the threat of future storms by accelerating the completion of a massive hurricane protection levee system.

Resilience, as used here, encompasses the capabilities of a community to rebound after a major disruptive event and is one component of a sustainable society. Within that general definition, Thomas Wilbanks observes that a resilient community "*anticipates* problems, opportunities, and potentials for surprises; *reduces* vulnerabilities related to development paths, socioeconomic conditions, and sensitivities to possible threats; *responds* effectively, fairly, and legitimately in the event of an emergency; and *recovers* rapidly, better, safer, and fairer" (Wilbanks 10, emphasis added). A human community, at least in theory, has the capabilities to learn from previous disruptive events, to adapt to potential threats, and marshal its capabilities to minimize future impacts—in other words, to become more resilient (Turner et al.). Adaptation is a fundamental capacity to enhance resilience; ultimately, resilience contributes to sustainability.

Social memory and public policy are two repositories for those lessons learned. Adger argues that after a disruptive event, social memories become "the growth points for renewal and reorganization of the social ecological system" (Adger et al. 1037). Endfield and others make the case that extreme events "can determine how people conceptualize risk and anticipate the impact of future events, and can be critical to the development of effective adaptation and response mechanisms" (Endfield et al. 239; Endfield). A resilient society embeds strategies and procedures for hazard mitigation and recovery into its public policy—development of improved hurricane forecasting to better anticipate storms, investments in levees to reduce future flood impacts, the creation of procedures and institutions to respond to and recover from disruptive hazards, along with a host of other statutory, regulatory, and institutional adaptations. While some adaptations have firm legal or institutional grounding, the social memory of major calamities tends to fade rather quickly, and the state of public preparedness erodes over a matter of months or years. Yet the interval between major disruptive events may be decades. Thus, perpetuating social memory becomes a necessary component of community resilience, and it is one that is frequently neglected (Colten and Sumpter).

Three areas of community life in New Orleans offer insight into the loss of social memory in terms of hurricane resilience: urban land use regulation and local hurricane protection structures, residential architecture, and evacuation. The first two reflect long-term personal and community investments in living spaces and how individuals and the city guided urban growth and home construction with hurricane-induced flooding in mind. The third is a much more flexible procedure, but evacuation practices reveal the enduring power of social memory embedded in institutional practices—even if it is no longer aligned with safety considerations.

Land Use and Hurricane Protection Structures

In its first two centuries of urban growth, New Orleans clung to the slightly higher and less flood-prone natural levee. In a city where topographic variation is largely invisible and the high ground is only about fifteen feet above sea level, urban growth took advantage of every inch of elevation to reduce the ever-present flood threat (Figure 1).

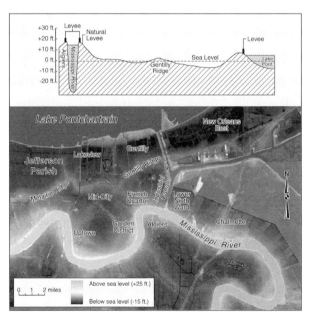

Figure 1: New Orleans and its topographic setting; cartography by Clifford Duplechin

Population growth prompted the installation of a city-wide drainage system after 1900, which enabled the extension of residential land northward across the low-lying wetlands fronting Lake Pontchartrain. By the time of the 1917 hurricane, lakefront development was modest, but leaders recognized this area as the new residential frontier. Flooding caused by storm surge and waves damaged 25,000 structures, and local authorities responded with a plan to armor the lakefront with a giant concrete seawall. Although it took nearly two decades to complete, by 1934 a nine-and-a-half-foot-high barrier stood between the new suburbs and the lake (USACE 1997; Colten 2009). The 1947 hurricane overtopped the levee and also caused extensive inundation in adjacent Jefferson Parish. As a consequence, local authorities secured congressional approval for the Corps of Engineers to construct an earthen levee to protect the postwar suburbs in Jefferson Parish's lakefront district (United States Congress 1946; United States Senate). Both levee-building efforts represent anticipation of future storms and constitute explicit steps to reduce susceptibility to future damages. While these constructions contributed towards strengthening resilience, their effectiveness had its limits. Local planning and zoning bodies took no steps to regulate development within the protected areas, nor was it a requirement to aim for flood-proof construction in order to provide adequate protection for residents in the inevitable event that a storm exceeded the limits of the levee's design.

When Hurricane Betsy made landfall in 1965, it flooded much of the eastern New Orleans area, portions of the new subdivisions behind the lakefront levee, and older sections near the city center (Figure 2). Flooding was widespread with approximately 15,000 houses suffering serious damage (USACE 1965). The absence of extensive development in eastern New Orleans prevented more extensive property damage. Displaying a typical pattern of response, the city turned to structural protection for areas of the city that had been inundated. Fortunately for local citizens, Congress was already reviewing a plan drawn up by the Corps of Engineers to build a more extensive hurricane protection levee system. The nation's legislative body swiftly approved appropriations to begin work on this massive undertaking. It called for larger levees around the highly susceptible eastern New Orleans district and around large portions of downstream St. Bernard Parish, and higher levees along the already armored lakefront neighborhoods. Once again, local authorities enacted no complementary land-use regulations or safe building codes to accompany the levee project (Burby; Colten 2009). The city would rely almost entirely on the levees.

In order to justify the expenditures for this costly project, the Corps had to prepare a cost-benefit analysis. The purpose of this exercise was to demonstrate to Congress that the dollar value of the project's benefits was greater than the expenditures. Since much of the territory to be protected was idle, low-value marshland, the federal engineers calculated the value of this property based on projected values assuming future residential development. Only by calculating prospective "damages to be prevented" was the Corps able to raise the benefits tabulation above the massive costs and ensure funding (Colten 2009).

With the assistance of local planners and zoning bodies, the Corps' projection was eventually realized. The New Orleans Planning Commission promptly approved numerous new subdivisions in the eastern New Orleans district (Figure 2) (New Orleans City Planning Commission).

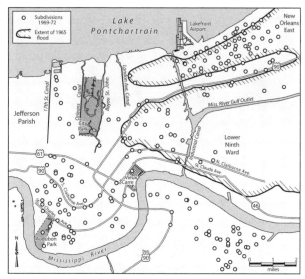

Figure 2: Flooding due to Hurricane Betsy in 1965 and subdivisions approved in the city in subsequent years; cartography by Clifford Duplechin

The city also worked with developers to allow the development of a massive expansion by New Orleans East—a private corporation with ambitious plans for the area receiving federal levee protection (New Orleans East). Before the local economic collapse of the mid-1980s, New Orleans East was one of the fastest growing areas in the metropolitan region (Burby). The city had authorized growth into the most flood-prone areas. Granted, the new levee system provided protection, but only for a 100-year storm. Resiliency diminished with every neighborhood that was approved in the lakefront neighborhoods and in adjacent St. Bernard Parish.

Residential Architecture

When French colonists arrived in New Orleans in the early eighteenth century, they hurriedly constructed rude structures using *poteaux en terre* technique—that is, they built directly on the ground. Regular river flooding prompted them to adopt construction techniques used by Caribbean planters (Figure 3).

Figure 3: Colonial era raised plantation house near the banks of the Mississippi
River in St. Bernard Parish; photo courtesy of the Louisiana State Museum

Raised houses, standing five to six feet above the ground, offered natural cooling
properties in the Caribbean and also proved highly adapted to the Mississippi River
floodplain by providing a flood-proof design (Edwards). Elevated houses became
an extremely common feature of New Orleans long before the development of
modern flood-proof standards that followed the passage of the National Flood
Insurance Act in 1968. Elevated structures found a place along the lakefront too,
where the modest houses of fishermen stood on wooden stilts several feet above
the wetlands (Figure 4).

Figure 4: Raised cottages over the wetlands facing Lake Pontchartrain, 1923;
photo courtesy of the New Orleans Public Library

Since only the wealthy could afford houses raised high on sturdy brick piers, the proliferation of working class housing in the late nineteenth century introduced a unique structure to the local landscape—the "double." The double shotgun house is one of the most common house types in New Orleans: it is a duplex, consisting of two long, narrow, parallel units (Figure 5).

Figure 5: Typical double shotgun house raised a few feet off the ground and one of the more common housing types built in the late nineteenth and early twentieth century as the city expanded into lower lying areas; photo courtesy of the New Orleans Public Library

Historically built of local cypress, the houses stand on brick piers about three feet above grade. This configuration enabled the residents to escape the more common flooding (and there had been no serious river flooding in the city since the late 1860s) and moreover, cypress wood is an exceptionally durable material that is resistant both to rot and termites. Thus, "doubles," and the single unit variant, the shotgun house, were adapted to local conditions. As the city expanded during the real estate and building boom of the 1920s, a new generation of raised houses appeared. Throughout the low-lying neighborhoods which sprang up shortly after the 1915 hurricane, bungalows with their living quarters above ground-level garages provided a new and distinctive New Orleans variation on a national style (Figure 6).

Figure 6: A raised bungalow in a low-lying area, 1929;
photo courtesy of the New Orleans Public Library

This arrangement offered flood-proofing for the living quarters and also a space for the new domestic necessity—the automobile. While builders erected many lower-standing doubles in the same time period, their elevated structure also offered a degree of resilient design.

Resilient construction declined in the years after World War II. As a clear reminder of the flooding potential, a 1947 hurricane subjected considerable areas near the city's northern lakefront to inundation. Despite this, the city expanded into the areas that had just endured high water. In terms of architecture, new building techniques favored slab-on-grade construction—that is, houses built on concrete foundations poured on the land surface. Between 1947 and 1950, the local levee board reported that contractors built 1,052 new houses in a flood-prone area that was at least three feet below sea level, and it projected that future storms would cause more damage to these ill-adapted residences (Orleans Levee District). Nonetheless, slab-on-grade construction predominated as the city sprawled into the lakefront districts in neighboring Jefferson Parish and eastern New Orleans (Figure 7).

Figure 7: Pontchartrain Park, a post-war subdivision for African Americans,
reflects the urban sprawl in the flood-prone lakeshore area, 1956;
photo courtesy of the New Orleans Public Library

Flooding caused by Betsy did little to re-inspire the use of raised housing. Even after the passage of the National Flood Insurance Act in 1968, the city took over two
decades to develop new codes and begin enforcing elevated construction. Flouting
the federal goal to minimize exposure to flooding, a federally financed housing
project in low-lying west-bank suburbs continued to employ slab-on-grade. Within
the greater New Orleans urban area, repeat flood payments reflected the lack of
flood-proof construction in new neighborhoods, and this became an issue of serious concern to the Federal Emergency Management Agency. It went so far as to
file a suit against Jefferson Parish in the 1980s to compel them to align their flood
protection practices with the federal programs' goals of reducing development in
flood-prone areas (Colten 2005). Across the urban area, over 150,000 new homes
were built between 1960 and 1990, and of those, 22,000 were erected in the repeatedly flooded eastern New Orleans district (Burby). Most new houses were slab-on-
grade, and they were in the lowest sections of the urban area. Resilience was essentially designed out of the city's architecture.

Local Evacuation

Evacuation from the areas most susceptible to hurricanes had become a well-
practiced response during the twentieth century, with the aim of reducing fatalities.
As forecasting improved, evacuation improved. In advance of the 1947 hurricane,
forecasters issued warnings for residents in coastal parishes and low-lying areas of
New Orleans to relocate to safer settings. Newspaper accounts indicate at least

23,000 individuals received sufficient notice to flee their homes (*NOTP* 19 Sep 1947, 1, 6, 9). At the time, individuals were dependent on their own vehicles or commercial bus services, and this ultimately limited the ability of some to escape. Evacuees from coastal communities and low-lying urban areas hastened to civic structures or hotels in the city which offered them temporary accommodation until they could return to their homes.

As Hurricane Betsy approached the gulf coast in September 1965, the Weather Bureau and local emergency responders in New Orleans tracked its movement. On September 7, based on the Weather Bureau's still imprecise forecasts, the New Orleans Corps of Engineers District office went on alert for possible emergency action two days before landfall. The Corps secured its fleet and dispatched emergency generators to facilities around the city (USACE 1965). The military evacuated aircraft from the Belle Chasse Air Station, and oil companies pulled their crews from the offshore rigs. Evacuees from the low-lying delta parishes crowded into New Orleans hotels, employing the traditional "vertical evacuation"—moving from the lowest areas to higher ground and up into hotels. The Department of Agriculture brought in emergency food to sustain 100,000 people in the city (*NOTP* 9 Sep 1965, 1). Public transit vehicles helped move evacuees to schools, government buildings, and military bases. Civil Defense plans had identified 187 sturdy, two to three-story school buildings scattered through the city as local shelters (Figure 8).

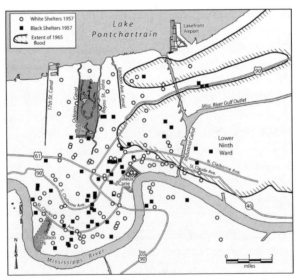

Figure 8: Areas flooded by Hurricane Betsy and schools designated as segregated shelters by the Civil Defense, 1957; cartography by Clifford Duplechin

No citizen had to travel a long distance to reach a facility and this minimized the need for lengthy advance notice. By the time the storm made landfall, overnight on September 9, the *Times Picayune* reported that the "largest single evacuation project in the region's history" had enabled nearly half a million people to move from the more vulnerable delta parishes and lakefront neighborhoods to shelters on higher ground (*NOTP* 9 Sep 1965, 1), with only two days' advance notice and no interstate highways to enable long-distance evacuation. New Orleans and the surrounding delta parishes had taken steps to secure offshore facilities, onshore government structures and equipment, and to evacuate a large number of people from the most exposed locations. The Corps estimated that about 90 percent of the population threatened by flooding (about 250,000 people according to its estimate) were successfully evacuated, which kept the death toll low (only 81) (USACE 1965). The core of the evacuation plan was local evacuation. It did employ a high degree of redundancy—multiple shelters within easy access of all residents—but this reflected effective resilience.

By the time of Katrina, several fundamental changes in the urban landscape had forced adjustments in the evacuation procedures. The massive hurricane protection levees reconfigured risk. If overtopped, they would serve as a bowl and capture flood waters to depths of up to 20 feet (Figure 9).

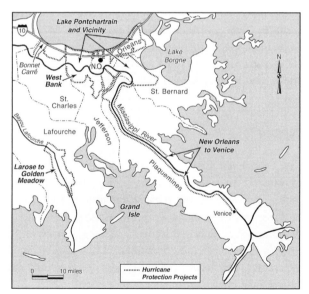

Figure 9: Expanded levee system that was designed after Betsy and was
still under construction in 2005; cartography by Clifford Duplechin

Thus, evacuation of populations into local schools was no longer a viable option. Limited-access expressways, largely completed after Betsy, offered three escape routes from the city. Evacuation plans had been adapted to align with these new realities, and relied on personal automobiles for long-distance evacuation beyond the city's encircling levees. Tested in 2004 when Hurricane Ivan threatened New Orleans, some 600,000 fled the city, only to become trapped in a massive traffic snarl. Planners modified the plan before Katrina to allow all interstate highway lanes to carry traffic away from the city (Colten 2009). This greatly improved traffic flow in 2005, but it did not account for the segments of the population with no access to personal transportation. For those individuals, the city opened the giant sports arena, the Superdome, as a shelter of last resort, and offered bus service for those needing assistance to get there. In the course of three days, at least 800,000 fled the city in their own vehicles with only moderate congestion, and approximately 10,000 huddled in the Superdome to ride out the storm (United States Congress 2006). Tens of thousands also remained in their residences with elderly or infirm family members or pets.

The shift to long-distance automobile evacuation reduced resiliency. This option requires more advance notice to accommodate the orderly departure of thousands of personal vehicles. It reduced the redundancy in terms of the number of shelters; however transportation provision for those without their own vehicles was inadequate. Furthermore, all three major interstate arteries pass over open water, susceptible to surge. If the evacuation is not completed far enough in advance, evacuees could be imperiled in the very act of fleeing the city. Indeed, Katrina destroyed several sections of the elevated highway that crosses the eastern end of Lake Pontchartrain and trapped some evacuees on isolated segments of the damaged expressway.

Losing Resilience

Those who lived through Betsy certainly retained vivid memories of that unforgettable storm and yet, nonetheless, the community's social memory suffered a degree of amnesia. There were frequent reminders from national and local sources about the impending threat of a devastating hurricane (*NOTP* 1990; McQuaid and Schleifstein; Sands; Fischeti). Modest storms in the intervening years prompted evacuations and pinpointed weaknesses in the hurricane preparation apparatus. Yet local leaders steered preparations away from the practices that offered the greatest resilience before Betsy.

The investment in and reliance on structural devices, levees in particular, reduced risk from moderate tropical cyclones. Yet they offered a false sense of security and literally obscured residents' view of the open water and wetlands surrounding the city. River levees had proven successful for over a century and compounded residents' ill-placed faith in the city's invulnerability to storm surge. After two centuries, builders and home buyers began to reject the effective practices of constructing houses elevated well above ground level. Civic authorities noted this situation in the early 1950s, but did nothing to make safe building practices a legal requirement in those areas previously vulnerable to floods, now newly ringed by levees. Finally, long-distance evacuation replaced redundant local shelters. Leaders effectively ignored the experience of history as reflected in these three dimensions of hurricane preparedness. As a consequence, the urban area was more susceptible to a lesser storm in 2005, even after massive expenditures to reduce the threat of flooding. Resilience lost does little to enhance sustainable coastal areas.

Works Cited

Adger, W. Neil et al. "Social-Ecological Resilience to Coastal Disasters." *Science* 308 (12 August 2005): 1036-39.

Burby, Raymond. "Hurricane Katrina and the Paradoxes of Government Disaster Policy." *Annals of the Association for Political and Social Sciences* 604 (2006): 171-91.

Colten, Craig E. *Perilous Place Powerful Storm: Hurricane Protection in Coastal Louisiana*. Jackson: University Press of Mississippi, 2009.

———. *An Unnatural Metropolis: Wresting New Orleans from Nature*. Baton Rouge: Louisiana State University Press, 2005.

Colten, Craig E. and Amy R. Sumpter. "Social Memory and Resilience in New Orleans." *Natural Hazards* 48.3 (2009): 355-64.

Colten, Craig E., Robert W. Kates, and Shirley B. Laska. *Community Resilience: Lessons from New Orleans and Hurricane Katrina*. Community and Regional Resilience Institute, 2008. Accessed 5 May 2011. <http://www.resilientus.org/library/FINAL_COLTEN_9-25-08_1223482263.pdf>.

Edwards, Jay D. *Louisiana's Remarkable French Vernacular Architecture, 1700-1900*. Baton Rouge: Louisiana State University, Department of Geography and Anthropology, 1998.

Endfield, Georgina. *Climate and Society in Colonial Mexico: A Study in Vulnerability*. Malden, MA: Blackwell, 2008.

Endfield, Georgina H., Isabel F. Tejedo, and Sarah L. O'Hara, "Conflict and Cooperation: Water, Floods, and Social Response in Guanajuato, Mexico." *Environmental History* 9.2 (2004): 221-47.

Fischeti, Mark. "Drowning New Orleans." *Scientific American* 285 (October 2001): 77-85.

Frey, William H. and Audrey Singer. *Katrina and Rita Impacts on Gulf Coast Populations: First Census Findings.* Brookings Institution, 2006. Accessed 5 May 2011. <http://www.brookings.edu/metro/pubs/20060607_hurricanes.pdf>.

Kates, Robert W., Craig E. Colten, Shirley B. Laska, and Stephen P. Leatherman. "Reconstruction of New Orleans after Hurricane Katrina: A Research Perspective." *Proceedings of the National Academy of Sciences* 103 (2006): 14653-60.

Liu, Amy and Allison Plyer. *An Overview of Greater New Orleans: From Recovery to Transformation.* Brookings Institution, 2010. Accessed 5 May 2011. <http://gnocdc.s3.amazonaws.com/NOIat5/Overview.pdf>.

Louisiana Department of Health and Hospitals (LDHH). *Enhancement of the U.S. Census Bureau 2006 Annual Population Estimates: Orleans Parish.* 2006. Accessed 5 May 2011. <http://www.popest.org/popestla2006/files/PopEst_Orleans_SurveyReport_01_11_07.pdf>.

McQuaid, John and Mark Schleifstein. "Washing Away." *New Orleans Times Picayune* 23-27 June 2002. Accessed 5 May 2011. <http://www.nola.com/hurricane/content.ssf?/washingaway/index.html>.

McKeithen, John. Congressional Testimony in US House of Representatives. *Hurricane Betsy Disaster of September 1965.* 89th Congress, 1st sess. 25 September 1965, 32.

National Hurricane Center. *Tropical Cyclone Report: Hurricane Katrina, 23-30 August 2005.* Accessed 5 May 2011. <http://www.nhc.noaa.gov/pdf/TCR-AL122005_Katrina.pdf>.

———. *Advisory No. 50, 11 PM, September 8.* 1965a. Accessed 5 May 2011. <http://www.nhc.noaa.gov/archive/storm_wallets/atlantic/atl1965/betsy/public/tcp30.gif>.

———. *Preliminary Report on Hurricane Betsy.* 1965b. Accessed 5 May 2011. <http://www.nhc.noaa.gov/archive/storm_wallets/atlantic/atl1965/betsy/prenhc/prelim08.gif>.

New Orleans, City of. *Situation Reports for New Orleans.* Produced between September 2005 and July 2006. Accessed December 2006. http://www.cityofno.com/portal.aspx?portal=1&tabid=66>.

New Orleans City Planning Commission. *Annual Reports.* New Orleans: New Orleans City Planning Commission, 1969-1970, 1970-1971, and 1971-1972.

New Orleans East, Inc. *A General Plan: New Orleans East, New Orleans, Louisiana.* New Orleans: New Orleans East, 1970.

NOTP, New Orleans Times Picayune. "N.O. Area Sitting Duck for Major Hurricane." 24 May 1990, 1-2.

———. "New Orleans Back in Business." 9 October 1965, 1.

———. "Lawless School Reopens Today." 5 October 1965, 1.

———. "Majority of Urban Areas Around N.O. have Power." 20 September 1965, 6.

————. "Betsy Cleanup." 19 September 1965, 13.

————. "Some Public and Catholic Schools to be Reopened." 16 September 1965, 29.

————. "Water, Sewerage Systems are Covered in Report." 16 September 1965, 24.

————. "Phone Service Being Restored." 15 September 1965, 8.

————. "Power, Phone, Transit Service." 12 September 1965, 1 and 9.

————. "All Big Stores to Open Today." 11 September 1965, 1.

————. "Thousands Flee Flood Threat as Hurricane Slams into N.O." 10 September 1965, 1.

————. "Nearly Half Million People Beat Betsy to Safe Areas. 10 September 1965, 1.

————. "Take Precautions, Quit Lakeshore for Safer Spot." 19 September 1947, 1.

————. "Delacroix Island Battens Down for Storm." 19 September 1947, 6.

————. "Residents fo Lower Coast Swarm into City for Safety." 19 September 1947, 9.

Orleans Levee District. *Report on Flood Control and Shore Erosion Protection of City Of New Orleans from Flood Waters of Lake Pontchartrain.* New Orleans: Orleans Levee District, 1950.

Sands, Thomas. Interview with Craig Colten. Rec. January 2006. Digital media.

Turner, B. L. et al. "A Framework for Vulnerability Analysis in Sustainability Science." *Proceedings of the National Academy of Sciences* 100.14 (2003): 8074-79.

US Army Corps of Engineers (USACE). *Interim Closure Structure at 17th Street Canal.* 2006a. http://www.mvn.usace.army.mil/hps/OEB09.htm. Accessed December 2010.

————. *Interim Closure Structure at London Avenue Canal.* 2006b. http://www.mvn.usace.army.mil/hps/PI_orleans_hpo.htm. Accessed January 2010.

————. *History of Hurricane Occurrences along Coastal Louisiana: 1997 Update.* New Orleans: U.S. Army Corps of Engineers, New Orleans District, 1997.

————. *Report on Hurricane Betsy, 8-11 September 1965. After Action Report.* New Orleans: U.S. Army Corps of Engineers, New Orleans District, 1965.

United States. Congress, House of Representatives, Lake Pontchartrain, Louisiana. *A Failure of Initiative: Final Report of the Select Bipartisan Committee to Investigate the Preparation for and Response to Hurricane Katrina.* Washington, DC: Government Printing Office, 2006.

————. *Letter from the Secretary of War, House Doc. 691.* 79th Congress, 2nd sess., 3 July 1946.

United States. Senate. Lake Pontchartrain, Louisiana, *Letter from the Secretary of the Army, Senate Doc. 139.* 81st Cong., 2nd sess. Washington, 2 February 1950.

Wilbanks, Thomas J. "Enhancing the Resilience of Communities to Natural and Other Hazards: What We Know and What We Can Do." *Natural Hazards Observer* 32.5 (2008): 10-11.

Dispersing Disaster:
The Deepwater Horizon, Ocean Conservation, and the Immateriality of Aliens

Stacy Alaimo

> The Gulf of Mexico is a very big ocean. The amount of volume of oil and dispersant we are putting into it is tiny in relation to the total water volume.
> (Tony Hayward, 14 May 2010)

> Deeply rooted in human culture is the attitude that the ocean is so vast, so resilient, it shouldn't matter how much we take out of it—or put into it.
> (Sylvia Earle 2009, 11-12)

Famed oceanographer Sylvia Earle alerts us to how the cultural conception of the ocean as vast and resilient enables humans not only to deposit such things as hazardous waste in the seas, but also to dispose of culpability, responsibility, and concern[1]. The 2010 Deepwater Horizon oil spill disaster in the Gulf of Mexico demonstrates this dynamic. It suggests how difficult it is to promote ocean conservation when industry, media, and culture imagine that the very scale of the ocean allows it to swallow up human-induced harms yet remain unscathed. Richard A. Kerr, in the article "A lot of Oil on the Loose, Not so Much to Be Found," states that although scientists are still wondering where all the oil that spewed into the Gulf of Mexico in 2010 ultimately went, the official US government report (drawn up by the Department of the Interior and the *National Oceanic and Atmospheric Administration*) claims that "75% has been cleaned up by Man or Mother Nature." Even more disturbing than the crass gender dichotomy here is that "nothing in the report supports that interpretation" (734). Moreover, while the report states that the oil "is biodegrading quickly," Kerr points out that no "documentation for that claim" is provided (735). Although the disaster itself was highly publicized—with photographs and videos of the oil spewing from the well, the unprecedented volumes of chemical dispersant being sprayed, and the gulf coast animals and shores covered with oil—now that the crisis is supposedly finished, having lived out its media lifespan, we are supposed to settle into a secure sense that the oil and the chemical

---

[1] The quote from Tony Hayward appears in "Tony Hayward, BP CEO: Gulf Oil Spill 'Relatively Tiny.'"

dispersants have disappeared: magically "cleaned up" by "Man" or "Mother Nature."

The discourse of "cleaning up" deserves critique, since it simplifies and ultimately trivializes the disaster, making it seem as if the "spill" can be readily remedied, just as one would wipe up spilled milk on a countertop. Yet, while such discursive critique remains vital for political contestation, the new materialist theories—such as those of Karen Barad and Nancy Tuana—would provoke scholars to attend to the "intra-actions" (Barad) or the "interactions" (Tuana) between ostensibly separate material and discursive domains. Barad's explanation of her "agential realist account," for example, refuses to endorse the gulf between nature and culture, discourse and the material world: "discursive practices are not human-based activities but specific material (re)configurations of the world through which boundaries, properties, and meanings are differently enacted" (183). Indeed, Barad's term "intra-action," "*signifies the mutual constitution of entangled agencies*," (emphasis in original, 33), which means that entities do not preexist their relations and that these relations are always part of the "world's radical aliveness" (33). The homey "cleanup" scenario, by contrast, assumes that the oil, the waters of the ocean, the creatures living in the ocean, "Man," and "Mother Nature" are somehow separate, stable, distinct entities. It assumes that an agent, "Man," or even "Mother Nature" can act in a direct, unimpeded manner in order to remedy the disaster, a disaster that is external to the agent. It removes human actions and knowledge practices from the turbulent sea of swirling material agencies, granting "Man" a secure position apart from Nature. It is crucial, however, in terms of ocean conservation, to recognize that the BP disaster unleashed substances that are themselves agential. They will have predicted and unpredicted effects on the plants, animals, and ecosystems of the Gulf waters, on the coastal areas, the open seas, and on who knows how many other habitats and living creatures, including humans. Some of these effects will be captured and documented by scientists, government agencies, and other groups, and some will not; the extent of our understanding will depend on economic, political, technological, ideological, and other factors. Some substances released by the BP disaster will interact with other toxins humans have dumped in the oceans, affecting sea life and coastal life. Some toxins will travel back to the land, carried by "seafood" consumed by humans and their domestic pets. It is impossible to simply "clean up" the BP disaster, since the disaster was not an isolated event, but instead, consists of ongoing, emergent, dynamic intra-actions. Thus, an ethical response to the disaster, in Barad's terms would not be "about right response to a radically exterior/ized other, but about responsibility and accountability for the lively relationalities of becoming of which we are a part" (393).

As the "new materialisms"—theories that account for the substance, significance, and agency of material forces—emerge across interdisciplinary theories of the humanities and social sciences, it is important to consider whether these paradigms can be extended to the cultural, philosophical, and political questions that the current crisis in ocean conservation demands. The ocean, especially the pelagic and benthic zones, poses particular challenges for the new materialisms, in that terrestrial humans have often found it more convenient to imagine that the seas are imaginary than to undertake the scientific, cultural, and political work necessary to trace substantial interconnections between human discourses, human practices, and marine habitats.

New Materialisms, Trans-Corporeality, and the Deep Sea

In the wake of the linguistic/discursive turn that has long dominated interdisciplinary critical theory, scholars in environmental philosophy, science studies, feminist theory, and other fields have been developing modes of analysis that can account for material, as well as discursive, agencies. The denial or bracketing of material agencies within predominant theoretical paradigms in the humanities has thwarted scholarly inquiries, especially in fields that focus on human corporeality, animals, or the physical world, such as feminist corporeal studies, disability studies, animal studies, science studies, and environmental studies. The environmental humanities must analyze the ideological, cultural, and discursive constructions of "nature," but, at the same time, study these conceptions in such a manner so as not to radically separate cultural formations from the actions, systems, and processes of the material world. It is crucial that the parameters and methodologies of the environmental humanities do not parallel the persistent yet distorting conceptual divide between "nature" and "culture." Bruno Latour, Donna Haraway, and Nancy Tuana, for example, have persuasively demonstrated that the very opposition between "nature" and "culture" is impossible to sustain, in that even a cursory examination of such phenomena as the hole in the ozone layer, the oncomouse, and Hurricane Katrina reveal no clear demarcation between these ostensibly separate domains.[2] As Tuana puts it, "witnessing the world through the eyes of Hurricane Katrina reveals that the social and the natural, nature and culture, the real and the constructed, are not dualisms we can responsibly embrace" (209).

While predominant theoretical paradigms since the linguistic turn have minimized or excluded material substances and forces from consideration, the consumerism of everyday life, especially in the mainstream of the United States, is predi-

---

[2]   See Latour 1993; Haraway 1989; and Tuana.

cated upon the manufactured ignorance,[3] denial, or dismissal of harmful material agencies circulating through food, water, clothing, furniture, cleaning products, and other substances. In *Bodily Natures: Science, Environment, and the Material Self*, I suggest that in the United States, if not in the rest of the industrialized world, one of the primary impediments to environmentalism is the pervasive denial of the unpredictable material agencies of the myriad xenobiotic substances that surround us. People routinely spread hazardous chemicals on their lawns, public restrooms spray toxic "air fresheners," and even food products and cosmetics riddled with carcinogenic substances are decorated with the ubiquitous "pink ribbon" of the Susan B. Komen Foundation, whose mission it is, ironically, to spread "awareness" of breast cancer. That pink ribbon epitomizes how successfully the semiotic streams of capitalist consumerism wash away any "awareness" of the harmful chemicals harbored by the products themselves. I argue that we need to cultivate a tangible sense of connection to the material world as well as an onto-epistemology that makes space for the unpredictable material agencies that will unfold as staggering amounts of xenobiotic substances become part of our bodies and environments. I propose that environmental movements and cultural theories situate themselves within "trans-corporeality," that is, the material interchanges across human bodies, animal bodies, and the wider material world. As the material self cannot be disentangled from networks that are simultaneously economic, political, cultural, scientific, and substantial, what was once the ostensibly bounded human subject finds herself in a swirling landscape of uncertainty where practices and actions that were once not even remotely ethical or political matters suddenly become so. Activists as well as everyday practitioners of environmental health, environmental justice, and climate change movements, work to reveal and reshape the flows of material agencies across regions, environments, animal bodies, and human bodies—even as global capitalism and the medical-industrial complex reassert a more convenient ideology of solidly bounded, individual consumers and benign, contained, passive products.

Trans-corporeality demands recognition of the material agencies that cut across human bodies, environments, and social and economic systems. As a mode of posthumanism, it occupies the outline of the human, only to dissolve corporeal boundaries by tracing how the substantial interchanges between bodies and places extend into global flows. In short, although trans-corporeality as an ethics and politics must trace the travels in toxins across geo-political boundaries, it nonetheless begins as an assessment of (post)human bodies in their own, local sites. Marine conservation, however, even for coastal peoples, demands some kind of reckoning

---

[3]   For more on "manufactured ignorance," see Proctor.

with distant and unknown pelagic and benthic zones. While environmental health and environmental justice movements foster embodied, participatory knowledge practices that contend with the local—regions, watersheds, neighborhoods, schools, playgrounds—ocean conservation movements must make sense of the open oceans that are beyond international law and the boundaries of nation states. Even though many coastal peoples have developed traditional ecological knowledges about the sea creatures that they fish, hunt, or otherwise encounter, and even though there are some amateur ocean scientists, such as those operators of dive boats and dolphin watches, who are especially knowledgeable about and committed to marine life, for most people sea life is "encountered" in highly mediated forms—such as films, photography, coffee table books, websites, and aquariums. Thus, the ocean eludes the feminist, environmentalist, and environmental justice models of ordinary experts, situated knowers, domestic carbon footprint analysts, and trans-corporeal subjects who take science into their own hands and conceive of environmentalism as a scientifically mediated but also immediate sort of practice.

Whether or not the new materialisms can extend their reach across the depths and breadths of the oceans may well depend, in part, on whether or not marine sciences have the funding, technology, and motivation to provide more data, images, accounts, and analyses of benthic and pelagic ocean environments. At this point in time, the massive global industries of marine dumping and marine extraction (of oil, minerals, and "seafood") dwarf the scientific studies of the marine ecosystems that are being radically altered or even destroyed before they can even be described. Furthermore, despite the rather long history of environmentalist concern for whales and dolphins, most ocean creatures and ocean ecologies float far beyond the habitat of Western environmental concern. Aptly, Tony Koslow begins his book, *The Silent Deep*, with a New Yorker cartoon in which one woman says to another, comfortably seated on a couch behind a coffee table, "I don't know why I don't care about the bottom of the ocean, but I don't" (n.p.).

It is well known to those who study histories of marine science and marine conservation movements that the vastness of the seas has buoyed the cultural conception of the ocean as so enormous, so powerful, so abundantly full of life that it is impervious to human harm. Kimberly C. Patton, in *The Sea Can Wash Away All Evils: Modern Marine Pollution and the Ancient Cathartic Ocean*, argues that several religious traditions have conceived of the sea as a place of purification: "many cultures have revered the sea, and at the same time they have made it to bear and to wash away whatever was construed as dangerous, dirty, or morally contaminating" (xi). Whether or not these religious beliefs have persisted, both the scale and the hazardous nature of what is dumped into the seas has changed, entirely, from ancient times. Nonetheless, contemporary global practices of dumping garbage, sewage, weapons, toxic chemicals, and radioactive waste assume that dispersing the substances or forces across the breadth and depth of the seas will make them disappear. Tony Hayward, the infamous representative of the BP Petroleum deep sea drilling disaster of 2010, merely echoes the common sentiment that the scale of ocean so fabulously diminishes any human action that all traces of human culpability ebb away into invisibility when he stated, "The Gulf of Mexico is a very big ocean. The amount of volume of oil and dispersant we are putting into it is tiny in relation to the total water volume" ("Tony Hayward" n.p.).

Ironically, despite the ostensible ability of the ocean to swallow all evils, BP chose to dump massive quantities of a chemical dispersant, Corexit, onto the massive oil spill, in order to make the oil disappear. If the primary reason for using dispersants is to help the oil biodegrade more quickly, three decades of research on dispersants has not demonstrated that this will be the case. Instead, the results of the research have been "mixed," showing "evidence for enhancement, inhibition, and no effect" (Johnson and Torrice 5).

It does not take a public relations expert to recognize that the oil itself—highly visible, easily photographed, as it coated coastal birds and even as it moved through the gulf waters in "plumes"—was not only physically sticky, but also visually, ethically, legally, and financially "sticky." Even as it was distributed by wind and tides, it stuck to BP, tagging them as the cause of dead birds, dead marine life, and blackened beaches. It is much more difficult, of course, to trace the short or long-term effects of the chemical dispersant, dumped in astonishing amounts. Carl Safina contends, "Personally, I think that the dispersants are a major strategy to hide the body, because we put the murderer in charge of the crime scene. But you can see it. You can see where the oil is concentrated at the surface, and then it is attacked, because they don't want the evidence, in my opinion" (2010: n.p.). The spectacle of the disaster is dispersed by the "Corexit" (which sounds like "corrects it"), but the environmental devastation is, most likely, exacerbated by the unprecedented volume of this chemical that was dumped into the living waters of the Gulf. As Arne Jernelöv states, "to disperse the oil may help protect birds and beaches, but it will increase the exposure of fishes, crustaceans, molluscs, and all other organisms that live and breath in water" (356). Joseph Romm, interviewing scientist Carys Mitchelmore, writes that the dispersants "can lead to far greater accumulation in living organisms of polycyclic aromatic hydrocarbons." Dispersing the oil into the depths of the gulf waters means that "subsurface creatures—from oysters to coral to larval eggs—that may never have had significant exposure to the oil are now going to get a double whammy, getting hit by the oil and by the dispersants." Worse, he continues, "the oil droplets are now in a form that looks like food (e.g. the same size as algae) to filter feeders like oysters" (n.p.).

Despite the vivid imagery of Romm's explanation, the BP disaster epitomizes how difficult it is to maintain a steady focus on the flow of harmful substances through the oceans. At what point do dispersed toxins become, in both senses of the word, "immaterial," both lacking in substance and devoid of consequence?

While it may be possible in the future to assess whether it was the BP Oil disaster that dealt the final blow to the endangered Kemp's Ridley sea turtle, pushing it over the edge into extinction, will it be scientifically possible to trace the broader effects of this disaster on hundreds of other sea creatures, coastal creatures, and ocean ecosystems a decade from now? When, if ever, do the oil and the dispersant stop affecting the living creatures and the environments that they have invaded? Moreover, will the political inclination and the funding be available to conduct the scientific studies necessary to assess the widespread, long-term damage to ocean life? How many scientists have already been bought up by BP, which has offered lucrative contracts to purchase their silence (Raines)? Is dispersing a disaster across the expanses of the globe, paradoxically, the supreme mode of political containment, as it makes it nearly impossible for environmentalists to trace causes, effects, and culpabilities? Even the word "disperse" bears a contradictory denotation, as the primary sense of "to cause to separate in different directions" may ultimately mean either to "disseminate" and "make known abroad" or, conversely, to "cause to disappear." Interestingly, the earliest sighting of the word "dispersant" in the *OED* dates only from 1944, when it was coined in the trade journal *The Petroleum Refiner*. Dispersants, in more than just an etymological sense, have their origins in the petroleum industry.

Ironically, BP's response to the Deepwater Horizon demonstrates that the deep waters of the ocean, indeed, lie beyond the horizons of concern. BP's public relations statement, available as a PDF on their website, entitled "Deepwater Horizon Spill Response: Dispersant Use," argues that the "intent of dispersant use is not to hide the oil," but instead to "minimize the damage that would be caused by floating oil" (British Petroleum 6). But the document also admits that the decision to use the dispersant favors the shore rather than the wider ocean. This rather repetitive document notes, in a few different bulleted statements, that dispersants "are used to break up the remaining surface oil before it can drift onto shore," that dispersants "principally are used to prevent oil sheen from reaching the shore," and that by "diluting and dispersing oil far from shore, they reduce the risk that oil will wash onto sensitive shoreline habitats" (British Petroleum 1). Clearly, the recovery efforts target the more readily accessible, more readily photographed region of the shore. And we would be remiss to ignore the fact that the damage to coastal areas is more easily calculated in terms of economic harm to the fishing and tourism industries that are seeking compensation for damages. While Gulf area residents and industries have already taken legal action against BP, the risk of lawsuits undertaken on behalf of the environments and organisms of the wider ocean is negligible. James Liszka, observing the parallels between the disasters of the Exxon Valdez and the Deepwater Horizon, states that "there is an over-emphasis by government and corporate officials on short-term damages to the tourism and fishing industries, with an under-emphasis on the long-run damages to the ecological commons" (23-24).

The open oceans and the deep seas epitomize a global "ecological commons," a commons that is not only ripe for the "tragedy" of which Garrett Hardin warned, but also in danger of being overlooked as well as overused. The BP document admits that a "decision to use dispersant involves balancing the risks to certain animals and plants at the water surface and in shoreline habitats against the potential risk to other organisms in the water column and seafloor" (British Petroleum 3). "Other organisms," nameless and invisible, are readily cast aside.

The Deep Sea as Alien Space

The lack of scientific understanding and public concern for these unnamed "organisms in the water column and seafloor" pose formidable challenges for new materialisms, as well as for conservation movements. Without an understanding of newly discovered or even as yet undiscovered marine creatures, and without basic knowledge of how various marine ecosystems function, it will be difficult for scientists to capture the effects of the BP oil disaster. Another type of challenge, however, is that many depictions of the deep seas in popular culture have contributed to the sense that the pelagic and benthic zones are not only distant and perhaps unknowable, but unreal. It is astounding to discover how many films, memoirs, scientific and popular accounts of the deep seas depict them as not only alien regions but as regions inhabited by aliens. Stefan Helmreich titles his anthropological study of microbial ocean science, *Alien Ocean*, in order "both to diagnose a scientific, social, and cultural imagination about the sea" and "to suggest the limits of representing this sea, for both oceanographers and social scientists" (xi). While the Deepwater Horizon disaster and the public relations campaigns that followed in its wake do underscore the representational limits of ocean environments, they also suggest how the metaphor of the ocean as an alien space may cast it, in the human imagination, as unreal, and thus immaterial. While Helmreich sees the figure of the alien as "a stranger who may be friend or foe," and "a channel for exchange between the oceanic and the human" (xi), I think the discourse of "alien oceans" is much more insidious, as it implies that ocean life is conjured by science fiction fantasies. It also suggests that even if the creatures of the deep do exist, they dwell in a place so distant, so foreign, that they are radically disconnected from terrestrial environments, processes, and flows. (Or, in more everyday parlance: aliens don't exist, but if they do, they don't matter, except at the box office or in Roswell, New Mexico.). Moreover, even as all postmodern disasters—whether they be natural, unnatural, or, more likely, a swirling mixture of the two—are depicted as media spectacles, ripe for consumption and then for forgetting, any disaster involving the oceans may be especially at risk of being dispersed into oblivion because of the underlying belief that ocean environments are not actually real places to begin with. The belief that it is possible to disperse harmful substances throughout living waters in such a way that they will no longer cause harm is problematic enough, but the fantasy that those very waters are not just difficult to encounter or understand but completely cut off from human terrestrial and coastal habitats and thus exist, exclusively, as highly imaginative flights of fancy, threatens to render ocean conservation efforts immaterial, irrelevant, unnecessary, even unthinkable.

Two films by James Cameron epitomize this strange desire to depict marine environments as alien. In the 1989 film, *The Abyss*, a lengthy Cold War drama, underwater oil rig workers and Navy SEALs are forced to work together to investigate an American ballistic missile submarine that has sunk. Despite the fact that the film is set in the ocean, with many long scenes of the characters moving through the water in dive suits and submersibles, the film does not feature any marine life. Sponges, corals, dolphins, sharks, rays, fish, eels, and jellyfish are absent. The seas are strangely devoid of organisms. Until, that is, near the end of the film, a vast civilization of alien creatures, living on the bottom of the sea, appears. (Their faces resemble that of E.T., while their bodies mix the shape of rays or angels with the gelatinous, watery substance of jellies.) These creatures, as well as their diaphanous city, significantly, are beautiful yet insubstantial, comprised of sheer, watery light. In this strange Cold War morality tale, the aliens decide, at the last minute, not to destroy the human race. The aliens, it turns out, have harnessed the power of water to create huge tidal waves that tower above the terrestrial humans, who scream and try to run away, only to be happily surprised when the waves abruptly halt at their apex—poised above the cities. The aliens change their collective mind because the protagonist, Bud (Ed Harris), gives his life to save them from the nuclear missile that the humans have dropped into their neighborhood. Not unlike Octavia Butler's Oankali, this alien species possesses plenty of evidence that humans deserve extermination: they show Bud a television montage of Hiroshima, Vietnam, the Holocaust, and other horrors. Significantly, however, in terms of marine environmentalism, the film erases existing ocean creatures and ecosystems, supplanting them with this imaginary alien civilization of glowing light. These beings are watery, ethereal, insubstantial, and ultimately, of course, not at all real. In fact, during the pivotal scene in which Bud encounters the aliens, they float above him like angels, accompanied by tinkling bells and choral music. Cut to the next scene: as the angel-like creatures fade out, the clouds fade in, visually merging the submerged aliens with the heavens as well as with outer space. Such transcendent moments divert our attention from the deep waters. While the hero of the tale is able to dismantle the nuclear missile that could have destroyed the aliens, there is no mention of the dispersed radioactivity and other pollutants that continually wreak their harms on marine creatures and environments that actually exist. The plot of the film securely contains the ongoing, destructive power of all the radioactive waste in the seas—from weapons disposal, the bombing of Pacific islands, and other sources—within that one small missile.

Cameron's more recent film, *Aliens of the Deep* (2005), documents deep ocean expeditions, in which he, along with marine biologists, astrobiologists, other scientists, an astronaut, and Russian Mir space station pilots explore the depths of the seas via submersible vehicles and rover cameras. Much of the film portrays engineering challenges and other technological dramas, highlighting the rather self-consciously enacted heroics of Cameron himself. Strangely, the viewer learns little about the ocean, the creatures the explorers encounter, or ocean conservation. When Cameron and planetary scientist Kevin Hand see a large gelatinous animal, Hand exclaims, "Look at that thing! That is absolutely unreal!" Cameron says, "Look at this thing. Look at this thing; It's just incredible. . . Beautiful." The film lingers here, letting us watch the entrancing gelatinous animal billowing like a translucent scarf. Despite this arresting and entrancing scene, the film's narrative and structure makes the ocean creatures themselves subservient to space exploration.

In one scene, Michael Henry, a marine animal physiologist, sitting underwater in a submersible and operating a roving robotic device, encounters an octopus. When the octopus grips the gripper of the device—a lovely moment in which the curious octopus, not the human, becomes the explorer—Michael says, "It was an extraordinary encounter. It was as though I got to shake hands with an alien." The connection between sea creatures and those creatures who may exist in space is not merely a metaphor or simile in this film, unfortunately. Indeed, the ocean is touted as the perfect practice arena for space explorers; marine biology is cast as a good starting point for astrobiology, and the samples from the ocean are just the "next best thing" for the planetary scientist to examine. The ethereal trumps the aqueous, the transcendent transcends the immanent. Marine biologist Dijanna Figuero's compelling and informative discussion of symbiosis in riftia, the giant tube worms, for example, is followed by a cut to Cameron telling Hand, "The *real* question is, can you imagine a colony of these on [Jupiter's moon] Europa?"

The film concludes with a scene that morphs from science to science fiction. Figuero, the marine biologist, descends in the same submersible used for the documentary scenes, but when she reaches out and presses her hand to the glass of the submersible in order to make contact with an octopus-like sea creature it becomes clear that the creature is an alien and the documentary is now science fiction. Even worse, despite the beauty and significance of the marine life forms that the film documents, the film's triumphant conclusion involves not only an imaginary alien encounter, but an obvious reference to Cameron's own imaginary civilization of oceanic aliens depicted in *The Abyss*. The concluding scene expands to reveal a vast, luminous, blue city, but contracts when we consider that we are no longer discovering living beings in the world's wide oceans but are trapped within Cameron's mind.[4]

If only to assure readers that equating deep sea creatures with aliens is not merely a James Cameron obsession, we should note the recent low-budget film *Monsters* (2010). *Monsters* (Gareth Edwards) features enormous glowing creatures that resemble octopi with extra legs, who have fallen from outer space and invaded the borderlands between Texas and Mexico. The unlikely and unnerving spectacle of the creatures moving through air as if it were water and somehow walking on long, thin tentacles suggests that the trope of ocean creatures as aliens has become entrenched in the popular imagination. The fact that these giant aquatic creatures survive on land and move in ways that defy the laws of physics goes unremarked. Interestingly, Edwards based the creatures on the same sort of research that has fascinated Cameron: "There has been scientific research that says Europa, one of the moons of Jupiter, is composed of ice with oceans beneath that ice that could sustain life, similar to what is here on Earth. It was based on that data that I thought of going with a creature that resembled something you'd see from the ocean. I immediately thought of either a crab or an octopus but ultimately thought something that resembled an octopus could be both scary but beautiful for audiences at the same time" (Wixson n.p.). The creatures, removed from their marine environments and the crises of ocean conservation, are conjured up to deliver the alternating pleasures of fear and beauty to human audiences.

Conclusion: Making Oceans Matter

Political struggles to protect the ocean from future disasters, as well as from the devastation wrought by the global dumping and pillaging, must contest the belief that all threats to the oceans become dispersed in its vast waters. Ocean conservation movements must also contest the odd but nonetheless pervasive sense that these waters are so vast and so unimaginable as to actually *be* unreal, alien, immaterial zones. Scholars in the environmental humanities and science studies can support ocean conservation movements both by contesting the cultural constructions that pose the pelagic and benthic zones as immaterial, and by tracing the substantial interchanges, interconnections, and flows between industrial practices, scientific knowledge, and cultural fabrications, as well as their effects on marine envi-

---

[4]   Cameron plans to transform deep sea habitats into science fiction at least one more time. He intends to descend 36,000 feet into the Mariana Trench to film footage for the sequel to *Avatar*.

ronments. This is a tall order, in that the very remoteness of the ocean depths seems to position them beyond the horizon of human concern.

Prominent ocean advocates Carl Safina and Sylvia Earle responded to the BP Oil disaster by attempting to bridge the disconnect between terrestrial humans and the ocean environments. Carl Safina, in a brief essay on the website CNN.com, argues that the BP disaster will harm the endangered Kemp's Ridley turtle, the bluefin tuna, and many migrating birds. He concludes his piece by lamenting our lack of palpable connection to the oceans: "Why do we fail to know this always, in our bones? The Gulf is not a thing unto itself. Neither is the oil eruption. We are all Gulf victims now" (2010: 2.). However sincere, this plea falls a bit flat, as it fails to construct or reveal substantial connections between terrestrial humans and the inhabitants of the seas. Moreover, by spotlighting familiar animals he fails to provoke concern for the many lesser known life forms that have been and will continue to be harmed by the disaster. Sylvia Earle, whose recent book is titled, *The World is Blue: How Our Fate and the Oceans Are One*, reasserts her argument that human life depends on the oceans during a PBS interview about the BP disaster: "We're all dependent on the sea. With every breath we take, every drop of water we drink, we're connected to the ocean. It doesn't matter whether you ever see the ocean or not. You're affected by it. You—your life depends on it" ("Gulf Coast Oil Spill"). Sadly, despite Earle's own passion for the ocean and its creatures, she finds it necessary to appeal to human self-interest.

While Safina relies on well-known animals to make his plea, Julia Whitty's essay in *Mother Jones*, "The BP Cover Up," is much bolder, introducing readers to less familiar creatures. The fourteen-page essay begins with photos of an octopod, lantern fish, and hatchetfish. What is most fascinating about this essay, however, is that Whitty begins by explaining how a vast collection of creatures migrates each day, "the mysterious movements of a vast community of organisms known as the deep scattering layer" (DSL):

> This aggregation of life forms was unknown until the 1920s, when early hydrographers mapping the ocean with sound encountered a daytime "seafloor" around 3,300 feet, which rose perplexingly toward the surface at night. Named for its echo-reflecting signature, the DSL was eventually recognized by marine biologists in 1948 to be layers of living creatures hiding on the cusp between perpetual twilight and darkness. (2)

The daily vertical movement of the DSL, as Koslow explains, is the "most massive animal migration on the planet, involving some hundreds of millions of tonnes of animals each day" (51). In the wake of the BP disaster, Whitty's attention to the DSL accomplishes at least two different things. Firstly, and more obviously, it demonstrates how deep sea ocean environments are enmeshed with the marine regions with which humans are more concerned, thus extending a conservationist ethos throughout the waters: "For the ocean, any loss of productivity in the deep scattering layer would be the biggest cataclysm of all—impoverishing the surface waters, depleting the coasts, cascading across the boundaries between ocean and land to denude both natural and human economies" (Whitty 14-15).

Secondly, and less obviously, Whitty's focus on the deep scattering layer may also counteract the perception that dispersing toxic substances renders them immaterial. The very term "scattering" parallels the action of "dispersing," but without the implication that what is scattered will dissolve or disappear. Ironically, as scientist Kelly Benoit-Bird explains to Whitty, "the DSL was a hot topic during the Cold War ... but only its acoustic properties, not its biological properties. American and Soviet navies wanted to know how to use its sound-reflecting properties to hide their submarines" (13). Once valued for its potential to shield military vessels, only originally detected because it scattered the pulses of the echo sounders, this "layer" is now valued by marine biologists as an essential part of the living oceans. Bruno Latour, arguing for the importance of the "flowing character" of science, explains that the scientific "chain" "leads toward what is invisible, because it is simply too far and too counterintuitive to be grasped directly" (2010: 122). The deep scattering layer of myriad living marine beings calls us to realize that what is, or was, invisible to human technologies of perception and knowledge should not be cast as simply alien or immaterial. Whitty provides a poetic rendition of the current moment in the flow of scientific understandings of the DSL:

> The emerging picture is one of an incalculably complex, finely tuned, and delicate interaction between predators and prey, chemistry and light, currents and water column, night and day. Some semblance of this spatial ballet, played in weightless three-dimensional darkness, has likely been part of the oceans since the oceans were brought to life: layers of life gathering in extremely high densities to feed or to avoid being eaten. (14)

Whitty explains that the oil plumes may trap the fish and invertebrates of the DSL above, below, or within them, preventing the migratory creatures from accessing their food, trapping them in a zone where they will surely be eaten, or enveloping them in poison. Her snapshot of this flow of living creatures captures the dynamic interplay between different organisms as well as between the organisms and their rich, dynamic, aquatic environments.

Scientific, journalistic, photographic, and artistic accounts of the effects of the BP oil and dispersant on specific ocean creatures, systems, and processes are crucial for rendering the ocean not as an alien or immaterial domain, but as a complex and dynamic environment that is significantly different from, yet interconnected with, the landscapes humans inhabit. Ocean creatures call on us to stretch our ability to conceptualize life itself, so that we will be more mindful and protective of environments that barely register on anthropocentric horizons as living places. Rather than projecting outer-space fantasies onto the deep seas, or using them as the proving ground for xenobiology, we need to remain—with the assistance of deep sea submersibles and other scientific and technological apparatuses—in the Gulf of Mexico and in other oceanic places around the globe in order to assess the avalanche of anthropogenic damages and to construct the most effective methods, policies, and social movements to minimize further harm. It may not be possible to cultivate meaningful modes of connection to oceanic creatures that are so far re-moved from human lives—a jellyfish is an unlikely candidate for a "companion species"[5]—but we could at least stop indulging in magical thinking that the ocean is impervious to human harm. Staying focused on the lives of specific animals—rather than evoking the ocean as a vast void—may be a way to foster movements for ocean conservation, movements that, despite the dreadful state of ocean health, are not at all inevitable. Minnie-Bruce Pratt's poem "Burning Water," composed as a response to the BP oil disaster, concludes with a memory of how small sting rays "winged" between her feet as she stood in the Gulf. After the BP disaster contami-nated the waters with oil and other toxic chemicals she wonders, "Now what will they eat?/ The connection between there and now not inevitable,/matter striking my mind, me trying to catch the spark."

Works Cited

The Abyss. Dir. James Cameron. Twentieth Century Fox, 1989.

Alaimo, Stacy. *Bodily Natures: Science, Environment, and the Material Self.* Bloomington: Indiana University Press, 2010.

Aliens of the Deep. Dir. James Cameron and Steven Quale. Walt Disney Pictures, 2005.

Barad, Karen. *Meeting the Universe Halfway: Quantum Physics and the Entan-glement of Matter and Meaning.* Durham, N.C.: Duke University Press, 2007.

British Petroleum. "Deepwater Horizon Spill Response: Dispersant Use." *British Petroleum website.* 19 June 2010. <http://www.bp.com/liveassets/bp_internet/globalbp/globalbp_uk_english/incident_response/STAGING/local_assets/downloads_pdfs/Dispersant_background_and_FAQs.pdf>.

Butler, Octavia. *Lilith's Brood.* New York: Warner, 2000.

[5]   See, of course, Haraway 2003.

Earle, Sylvia. *The World is Blue: How Our Fate and the Oceans are One.* Washington, DC: National Geographic Society, 2009.

"Gulf Coast Oil Spill Adds 'Insult to Injuries' for Ocean's Health." PBS News Hour. Transcript of Judy Woodruff's Interview with Sylvia Earle. Aired 5 May 2010.   <http://www.pbs.org/newshour/bb/environment/jan-june10/oil2_05-05.html>.

Haraway, Donna. *The Companion Species Manifesto: Dogs, People, and Significant Otherness.* Chicago: Prickly Paradigm, 2003.

———. *Modest_Witness@Second_Millennium.FemaleMan©_Meets_OncoMouse™.* New York: Routledge, 1989.

Helmreich, Stefan. *Alien Ocean: Anthropological Voyages in Microbial Seas.* Berkeley: University of California Press, 2009.

Jernelöv, Arne. "The Threats from Oil Spills: Now, Then, and in the Future." *AMBIO: A Journal of the Human Environment* 39 (2010): 353-66.

Johnson, Jeff and Michael Torrice. "BP's Ever-Growing Spill." *Chemical and Engineering News* 88.24 (2010):15-24.

Kerr, Richard A. "A Lot of Oil on the Loose, Not so Much to Be Found." *Science* 329 (2010): 734-35.

Koslow, Tony. *The Silent Deep: The Discovery, Ecology, and Conservation of the Deep Sea.* Chicago: University of Chicago Press, 2007.

Latour, Bruno. *On the Modern Cult of the Factish Gods.* Durham: Duke University Press, 2010.

———. *We Have Never Been Modern.* Trans. Catherine Porter. Cambridge: Harvard University Press, 1993.

Liszka, James. "Lessons from the Exxon Valdez Oil Spill: A Case Study in Retributive and Corrective Justice for Harm to the Environment." *Ethics & the Environment* 15.2 (2010): 1-30.

Monsters. Dir. Gareth Edwards. Vertigo Films, 2010.

Patton, Kimberley C. *The Sea Can Wash Away All Evils: Modern Marine Pollution and the Ancient Cathartic Ocean.* New York: Columbia University Press, 2006.

Pratt, Minnie Bruce. "Burning Waters" Poets for Living Waters. Accessed 15 January 2011. <http://poetsgulfcoast.wordpress.com/>.

Proctor, Robert N. *Cancer Wars: How Politics Shapes What We Know and Don't Know About Cancer.* New York: Basic Books, 1995.

Raines, Ben. "BP Buys up Gulf Scientists for Legal Defense, Roiling Academic Community." *Alabama Live* 16 July 2010. <http://blog.al.com/live/2010/07/bp_buys_up_gulf_scientists_for.html>.

Romm, Joseph. "Is BP's Remedy for the Oil Spill Only Making it Worse?" *Salon.com* 10 May 2010. <http://www.salon.com/news/feature/2010/05/05/oil_dispersants_poisoning_gulf>.

Safina, Carl. "The Oil Spill's Unseen Culprits, Victims." *TED: Ideas Worth Spreading,* video. Accessed 20 January 2011. <http://www.ted.com/talks/carl_safina_the_oil_spills_unseen_culprits_victims.html>.

———. "We Are All Gulf Victims Now." *CNN Opinion* 3 June 2010: 1-2. <http://articles.cnn.com/2010-06-03/opinion/safina.gulf.wildlife.impact_1_bluefin-tuna-oil-disaster-turtle?_s=PM:OPINION>.

"Tony Hayward, BP CEO: Gulf Oil Spill 'Relatively Tiny.'" *Huffington Post.* Accessed 19 January 2010. <http://www.huffingtonpost.com/2010/05/14/bp-ceo-gulf-oil-spill-rel_n_576215.html>.

Tuana, Nancy. "Viscous Porosity." *Material Feminisms.* Ed. Stacy Alaimo and Susan Hekman. Bloomington: Indiana University Press, 2008. 188-213.

Whitty. Julia. "The BP Cover Up." *Mother Jones* September/October 2010. Accessed 10 December 2010. <http://motherjones.com/print/71351>.

Wixson, Heather. "Exclusive: Director Gareth Edwards talks Monsters." *DreadCentral.com.* Accessed 24 January 2011 <http://www.dreadcentral.com/news/39359/exclusive-director-gareth-edwards-talks-monsters>.

Notes on Contributors

Stacy Alaimo is a professor of English at the University of Texas at Arlington where she co-chairs the President's Sustainability Committee. She received her PhD from the University of Illinois and has published widely on literature, film, environmental art and architecture, performance art, feminist theory and nature, environmental pedagogy, gender and climate change, and the science and culture of "queer" animals. Among her publications are *Undomesticated Ground: Recasting Nature as Feminist Space* (2000), *Material Feminisms*, edited with Susan J. Hekman (2008), and *Bodily Natures: Science, Environment and the Material Self* (2010). Her new book project is tentatively titled *Sea Creatures and the Limits of Animal Studies: Science, Aesthetics, Ethics*.

Craig E. Colten is the Carl O. Sauer Professor of Geography at Louisiana State University. Before coming to Louisiana he earned a PhD from Syracuse University, worked for the state of Illinois, and for a private consulting firm in Washington, DC. His research focused on urban hazards and lead to the publication of *An Unnatural Metropolis: Wresting New Orleans from Nature before Hurricane Katrina* (2005). In the wake of that storm, Colten provided insight about New Orleans to the BBC, CNN, the *New York Times*, and other international media outlets. More recently he published *Perilous Place, Powerful Storms: Hurricane Protection in Coastal Louisiana* (2009). Colten serves as a research associate with the Community and Regional Resilience Institute based at the Oak Ridge National Laboratory.

Lawrence Culver is an associate professor in the Department of History at Utah State University. He received his PhD from the University of California, Los Angeles, in 2004. His dissertation received the 2005 Rachel Carson Prize for best dissertation from the American Society for Environmental History. Culver is the author of *The Frontier of Leisure: Southern California and the Shaping of Modern America* (2010). In 2010 Culver was in residence as a fellow at the Rachel Carson Center in Munich to work on his book project, *Manifest Disaster: Climate and the Making of America*. Culver is a chair of the Alumni Association of the Rachel Carson Center.

Heike Egner is a professor of human geography at the Alpen-Adria University in Klagenfurt, Austria. Her research interests are in the relationships between humans and the environment, and in cultures of risk in modern societies. Before joining the faculty at Klagenfurt, Heike Egner was co-director of a project titled "Communicating Disaster" at the Center for Interdisciplinary Research in Bielefeld. Egner has taught at the universities of Mainz, Frankfurt am Main, Kassel, Vienna, Munich, and Innsbruck. In 2009 and 2010 she was a fellow at the Rachel Carson Center in Munich. She is the author of *Gesellschaft, Mensch, Umwelt—beobachtet* (2007), and *Geographische Risikoforschung: Zur Konstruktion verräumlichter Risiken und Sicherheiten* (2010).

Andreas Falke holds a chair in international studies (with a focus on Anglo-American cultures) at Friedrich Alexander University Erlangen-Nuremberg, and he is director of the German-American Institute in Nuremberg. His areas of speciali-zation include trade policy, the WTO, transatlantic relations, the political and eco-nomic system in India, and the domestic, foreign, trade, and economic policy of the United States. Falke was educated at Georg August University Göttingen, where he completed his dissertation in 1985 and received a post-doctoral degree (*Habilita-tion*) in 1996. Between 1983 and 2002 he was an advisor for social and economic public affairs programs and the principal economic specialist at the US Embassy in Bonn and Berlin.

Andrew C. Isenberg is a professor of history at Temple University in Philadel-phia, with an interest in environmental history, the North American West and the encounters between Native Americans and Euroamericans. He received his PhD in history from Northwestern University. Isenberg is the author of *The Destruction of the Bison: An Environmental History, 1750-1920* (2000), *Mining California: An Ecological History* (2005), and the editor of *The Nature of Cities: Culture, Land-scape, and Urban Space* (2006). In 2010 Isenberg was a fellow at the Rachel Car-son Center in Munich. Among other projects, Andrew Isenberg is currently editing the *Oxford Handbook of Environmental History*.

Sherry Johnson is an associate professor of Latin American and Caribbean history at Florida International University in Miami. Her research and teaching interests include Cuba and the Caribbean, environment and climate change, natural disas-ters, medicine, women and gender, and social history. She has worked on eigh-teenth century smallpox epidemics in the Hispanic Caribbean, domestic violence in Cuba, nineteenth century Cuban intellectuals, and hurricanes and climate change in the Americas. Johnson was a fellow at the Rachel Carson Center in Munich in 2010. Her publications include *The Social Transformation of Eighteenth Century Cuba* (2001) and *Climate, Catastrophe, and Crisis in Cuba and the Atlantic World in the Age of Revolution* (2011).

Christof Mauch is director of the Rachel Carson Center for Environment and So-ciety and professor of American cultural history and transatlantic relations at LMU Munich. Before coming to Munich he was director of the German Historical Insti-tute in Washington, DC. His recent publications include *Uncertain Environments: Natural Hazards, Risk, and Insurance in Historical Perspective*, edited with Uwe Lübken (2011), *The United States and Germany During the Twentieth Century: Competition and Convergence*, edited with Kiran Patel (2010), and *Natural Disas-*

ters, Cultural Responses: Case Studies Toward a Global Environmental History, edited with Christian Pfister (2009).

Sylvia Mayer is professor of American studies and Anglophone literatures and cultures and director of the Bayreuth Institute for American Studies (BIFAS) at the University of Bayreuth. Her major fields of research are ecologically oriented literary and cultural studies, and African American literature. Her publications include monographs on Toni Morrison's novels and on the environmental ethical dimension of New England regionalist writing. She has edited and co-edited several volumes of criticism, among them *Restoring the Connection to the Natural World: Essays on the African American Environmental Imagination* (2003), and *Literature, Culture, Environment: Positioning Ecocriticism* (2005).

Alexa Weik von Mossner is a postdoctoral fellow and lecturer in American literature and culture at the University of Fribourg in Switzerland and an associate of the Rachel Carson Center in Munich. She received her PhD in literature from the University of California, San Diego with a dissertation titled *Beyond the Nation: American Expatriate Writers and the Process of Cosmopolitanism* (2008). Her essays have appeared in journals such as *English Studies*, *Environmental Communication*, *Ecozona*, and in the *Journal of Commonwealth and Postcolonial Studies*. Weik von Mossner is currently working on a book about the imagination of global environmental risk in American popular media narratives.

Gordon Winder was appointed professor of economic geography at LMU Munich in 2011. Prior to that he was a research fellow at the Rachel Carson Center in Munich and an honorary research fellow at the University of Auckland's School of Environment. His research interest is in urban environments, historical experiences of industrialization, natural disasters and media, and globalization. Winder gained his PhD from the University of Toronto, and he has taught in Toronto, Auckland, and Munich. His book *Harvesting Networks: Historical Geographies of the American Reaper (1830-1910)* will be published by Ashgate in April 2012.